Daniel McAlpine and The Bitter Pit

Douglas G. Parbery

Daniel McAlpine and The Bitter Pit

Springer

APPS

Douglas G. Parbery
University of Melbourne
Blackburn, VIC, Australia

ISBN 978-3-319-09551-6 ISBN 978-3-319-09552-3 (eBook)
DOI 10.1007/978-3-319-09552-3
Springer Cham Heidelberg New York Dordrecht London

Library of Congress Control Number: 2014953915

Printed on acid-free paper

Springer is part of Springer Science+Business Media (www.springer.com)

Preface

DANIEL McALPINE
PIONEER PLANT PATHOLOGIST
1849 to 1932

In brief notes on famous Australian botanists in 1949, the late Dr Jim Willis, then editor of *The Victorian Naturalist*, ended his entry on Daniel McAlpine with the following remark:

> *Like so many genii, he just retired to fade out of public notice and die in obscurity. I have sifted through the proceeds of publications that one would expect to be loud in praising his public worth, but the only tribute I can find is in* the Australasian Journal of Pharmacy *for October and November 1932.*

It was a different story among the press and trade journals and in private correspondence with his family, but from the time of his death until the formation of the Australasian Plant Pathology Society in 1970, his worth was little appreciated other than by those whose own work led them into the world of this remarkable man.

When I joined the staff of the School of Botany at the University of Melbourne at the beginning of 1962, I gradually became aware that Daniel McAlpine's reputation within the School of Botany seemed not to be appreciated, and over time through discussions with Dr Ethel McLennan, I gained the impression that he had been deemed unsuccessful in his investigation of the cause and control of a serious problem of pome fruits, called 'bitter pit'.

In the early part of the twentieth century, revenue from apple production constituted a substantial proportion of the incomes of the southeastern Australian states. Consequently, the advent of bitter pit, which at times wiped out whole crops of apples and which, in some years, reduced the national apple crop by 50 % or more, and destroyed whole consignments of apple during the long voyages to Britain, was of national concern. This was serious enough to involve prime ministers such as Andrew Fisher, Joseph Cook and Billy Hughes to take nationwide action to understand and reduce the threat.

In 2010, Drs Peter Merriman and Peter Ridland were asked to sort through a pile of old files in the library at the Knoxfield Centre of the Victorian Department of Primary Industries, in preparation for its closure. Among the clutter, two significant files were found. One was the long believed lost manuscript of Charles French Snr's sixth volume of *Destructive Insects of Victoria*, since published in 2013. The second consisted of over a thousand documents detailing the lead into, the course, and the aftermath of the bitter pit investigation, which demonstrate McAlpine's substantial success in reducing the problem, but resulted in his ignominious retirement. The scientific knowledge required to elucidate the cause of bitter pit was not available until 40 years after the conclusion of the investigation. The consequence was that a reasonable understanding of the cause of this complex mineral deficiency did not emerge until the 1970s. Even without such knowledge, McAlpine was edging towards understanding the fundamental physiological process and physical modifiers involved, but in a vacuum of knowledge about minor and trace nutrients and the physiology of maturing and ripening fruit; he was denied the missing pieces of the puzzle.

He confirmed that neither infectious agents nor poisons were the cause. He realized that water availability was involved, but was confused by the observation that both deficiency and excess of water often led to the same result. The discovery of these files has enabled me to complete a fuller biography of this remarkable man.

Than was previously possible and to demonstrate that even though he did not regard himself as a plant physiologist, and was reluctant to be involved in the investigation, his success in helping the horticultural industry was equivalent to his success in controlling many fungal diseases.

Knowledge of McAlpine's contribution to mycology and plant pathology is generally known and appreciated because it is readily available in the literature. While the results of his investigation of the control of bitter pit are recorded in his Annual Reports published between 1912 and 1916, the story of the events leading up to the investigation, the harassment he endured during the investigation and the unjust and cruel treatment meted out to him at the end of it is not. It has mouldered away in forgotten files for over 90 years in out-of-the-way repositories. For these reasons, the greatest portion of this book deals with McAlpine's investigation of the bitter pit, which, while benefiting the apple trade, did great harm to him.

Blackburn, VIC, Australia Douglas G. Parbery

Acknowledgments

I have received wonderful help from many friends and colleagues over the three years of writing Daniel McAlpine's story for which I am deeply grateful and wish to thank the following.

Dr. Peter Merriman and Dr. Vyma Beilharz were instrumental in drawing me into the project and have helped me considerably throughout. Advice and assistance has also been gratefully received from Drs. Jacquie Edwards, Robert Holmes, Greg Johnson, Tom May, Ian Pascoe, Peter Ridland, and John Walker. Vyma and Jacquie and Peter Merriman provided valuable comments and corrections of early drafts while Vyma and Jacquie lent me copies of McAlpine's books, found the original portraits of Brittlebank and McAlpine, and helped me in many others ways, as they have done over several decades.

Ms. Helen Taylor, archivist at Heriot-Watt University, and Dr. G.D. Eddie, archivist at the University of Edinburgh, provided information relating to McAlpine's activities in Edinburgh from 1875 to 1884. Similarly, I am grateful for the generous assistance given by the librarians and archivists at the National Library of Scotland, Edinburgh; the Baillieu Library of the University of Melbourne; the State Library of Victoria, Melbourne; and the Library of the National Herbarium, Royal Botanical Gardens, Melbourne.

I wish also to acknowledge the information I gleaned through conversations with the late Dr. Ethel McLennan over many years, and from discussions with, the late Isobel Cookson, both of whom experienced life in the School of Botany at Melbourne University during the Ewart era. The late Erica Wedge (nee McAlpine) gathered as much as possible of her father's correspondence which she placed in archives together with her recollections of his life and work. She also shared her knowledge over cups of tea with many of us, while expressing her hope that one day the events surrounding her father's trials and tribulations during the investigation of the cause and control of bitter pit in apples would be told. Little did I expect at the time that the privilege would fall to me.

I thank Drs. Eileen Davison, Peter Merriman, and Greg Johnson for their efforts towards having McAlpine's story presented in book form with Springer Publications. I also thank Zuzana Bemhart, Mariska van der Stigchel, and Ms. S. Rajeswari among other staff at Springer for their wonderful support and enthusiasm, and their tolerance of the time taken by me to return corrected proofs.

Finally and most importantly, I wish to acknowledge my wife Betty for her patient tolerance of the three years of her being a book-widow, while I locked myself away with Daniel McAlpine, and for her encouraging support and care.

Thank you,
Daug Parbery

Contents

Part I Origin and Development of a Pioneer Plant Pathologist

1 Introduction 3
1.1 Greatness Forgotten 3
1.2 Documents and Letters 5
References 6

2 Scotland & England 7
2.1 Scottish Origin and Early Years 7
2.2 Fascination with Geology 7
2.3 Undergraduates in London 9
2.4 Teaching Atlases and English Texts 10
2.5 Academia in Edinburgh 14
References 15

3 Highly Educated Non-graduate 17
3.1 Proof of Appointments in Edinburgh 18
3.2 Academic Career 19
3.3 Search for Greener Fields 21
References 22

4 Australia Felix 25
4.1 Establishment in Australia 25
4.2 Ormond College 26
4.3 Auld Po-Tarsh: Victorian College of Pharmacy 28
4.4 Research and Other Occupations 1884–1890 29
4.5 A New Vocation Beckons 30
4.6 Natural Sciences in Nineteenth Century Europe 31
4.7 Australia 1788–1890 35
4.8 Formal Courses in Plant Pathology 37
References 39

5 Vegetable Pathologist 41
5.1 Devastating Epidemics 41
5.2 Founding Australian Plant Pathology 45
5.3 Renowned Mycologist 48
5.4 McAlpine's Training in Mycology 51
5.5 Hallmarks of Rigour and Accuracy 53
5.6 Six Australian Books 54
5.7 Extra Mural Activities 61
References 64

6 Twenty Years of Plant Pathology in Australia 1890–1910 67
6.1 In McAlpine's Own Words 67
6.2 What Constitutes Plant Pathology 68
6.3 Its Aims and Objectives 68
6.4 Means Provided for Investigation and Research 69
6.5 Some of the Work Accomplished 70
6.6 Problems Awaiting Solution 72
6.7 How This Study Has Benefited the Producer 73
6.8 Conclusion 74
6.9 International Recognition 75
References 77

Part II The Bitter Pit Investigation

7 Bitter Pit 81
7.1 History of Research into Bitter Pit 81
7.2 Bitter Pit in Australia 85
7.3 The Economic Importance of Apples 86
7.4 An Increasing Problem 1886–1910 87
7.5 A National Investigation Proposed 88
7.6 The Perth Conference 90
7.7 The Ministers Respond 90
7.8 The Reluctant Appointee 93
References 93

8 The Poison Theory 97
8.1 Codlin Moth, Lead Arsenate and Bitter Pit 97
8.2 Delay, Uncertainty and Objection 99
8.3 *Penang* – Investigative Journalist 100
8.4 The Minister and the Poison Theory 103
8.5 Test Committee Fails Ewart and White 104
References 105

9 A National Investigation 107
9.1 Government in Australia 107
9.2 Appointment of McAlpine 108
9.3 Commonwealth Advisory Committee on Bitter Pit 109
9.4 Travel, Distance and Time 110
References 111

10 The First Year 1911–1912 113
10.1 Establishing Starting Points 113
10.2 The Australian Apple Industry 1900–1911 113
10.3 Advisory Committee's First Report 118
10.4 The First Year's Accounts 121
References 121

11 Serious Stumbling Blocks 123
11.1 The Irony of Preparation 123
11.2 Ewart Corrupts Advisory Committee 123
11.3 Distribution of the First Progress Report 129
References 131

12 The Second Year 1912–1913 133
12.1 The Second Progress Report 133
12.2 Monthly Reports 135
12.3 Long Distance Transport of Meat, Fruit and Vegetables 138
12.4 The Voyage of *SS Somerset* 140
12.5 The Advisory Committee's Second Report 141
References 143

13 The Third Year 1913–1914 145
13.1 Monthly Reports 145
13.2 The Third Progress Report 150
References 151

14 The Fourth Year of the Investigation 1914–1915 153
14.1 Royal Commission 153
14.2 The Fourth Progress Report 156
14.3 The Advisory Committee's Penultimate Report 157
14.4 McAlpine's Letter to the Premier 159
References 160

15 Fifth Year of the Investigation 1915–1916 161
15.1 The Uncertainties of 1915–1916 161
15.2 Extension into a Fifth Year 163
15.3 The Final Advisory Committee Report 165
References 166

16 Appraisal of McAlpine's Success 169
16.1 Physiological Disorders 169
16.2 Quantitative Evidence of Success 171
16.3 Qualitative Evidence – The National View 173
16.4 The Prophet in His Own Land 174
16.5 International Comment 176
References 178

17 A Century Later 181
17.1 The Test of Time 181
17.2 Circling Around the Cause 182
References 186

Part III Twilight of a Career

18 Personal Interactions 191
18.1 Alfred Ewart 191
18.2 Ewart's Antagonism to McAlpine 191
18.3 McAlpine and the Advisory Committee 198
18.3.1 Divergent Objectives 199
18.3.2 Knowledge of Mineral Nutrients 200
18.4 Daniel McAlpine and Charles Brittlebank 200
References 203

19 The Fruitless Years 205
19.1 Termination and Withdrawal 205
19.2 Forgotten Gentleman's Agreement 205
19.3 Attempts to Continue Working – 1917 206
19.4 *Invercliffe* and Retreat into Retirement 212
19.5 The Espiranto Dictionary 214
19.6 Applied Evolution 214
References 217

20 A Reputation Re-established 219
20.1 Rekindling Appreciation 219
20.2 The Australasian Plant Pathology Society 220
20.3 The Daniel McAlpine Memorial Lectures 222
20.4 The Daniel McAlpine Outstanding Achievement Award 223
20.5 The Daniel McAlpine Medal 223
20.6 Wide Influence and Many Talents 224
References 225

Appendices ... 227
Appendix 1 ... 227
Biographical Notes on Some Who Touched on or Who Influenced McAlpine's Career ... 227
Appendix 2 ... 248
Appendix 3 ... 251
Daniel McAlpine Memorial Lecturers ... 251

List of Figures

Fig. 1.1 Herbert Hice Whetzel (1877–1944) 4

Fig. 2.1 Frontice Piece from McAlpine's Botanical Atlas Vol. 1 (1882) 12
Fig. 2.2 The Sweet Pea Botanical Atlas Vol 1, Plate XXI (1882) 13
Fig. 2.3 I infection hypha of *Phytophthora infestans* copied from De Bary. Botanical Atlas Vol. 2, Plate XII, Fig. 8 (1883) 14

Fig. 4.1 *Auld Po-Tarsh*, explaining some intricacy of botany to students on a field excursion 29
Fig. 4.2 McAlpine's life history of *Phytophthora infestans.* Based on de Bary. Botanical Atlas Vol. 2, Plate XII (1883) 33

Fig. 5.1 Map of the wheat growing areas of Australia. The unshaded eastern coastal strip was abandoned during the late nineteenth century. Ted Henzell (2007) *Australian Agriculture – Its History and Challenges*, Fig. 1.8, p 33 42
Fig. 5.2 Delegates at the Intercolonial Rust Conference. Sydney, June 1891. Standing; D. McAlpine (Vic.), W. Farrer (NSW), H. Kelly (SA), G. Inglis (SA), N. Cobb (NSW), H.C.L. Anderson (NSW), Seated: M^{c}. Knight (Vic), F.W. Shelton (Qld), Hon. Sydney Smith (NSW), P. McLean (Qld), A.N. Pearson (Vic) 44
Fig. 5.3 Black Leg of Cauliflower by C.C. Brittlebank. Vegetable Pathologist's Report, Department of Agriculture Victoria for 1900 46
Fig. 5.4 Illustrations of *Peziza* and *Penicillium. Botanical Atlas* Vol. 2, Plate XIV (1883) 52
Fig. 5.5 Charles Clifton Brittlebank (1862–1945) 55

Fig. 5.6 Life Cycle of *Puccinia graminis*. Botanical Atlas Vol. 2, Plate XVI (1883). Composed from Dodel-Port, De Bary and Tulasne 57
Fig. 5.7 Henry Tryon (1856–1943) c.1920 60
Fig. 5.8 British Association visitors on a Bay-side excursion with McAlpine in 1914 63

Fig. 6.1 McAlpine, Departmental Officers and Farmers inspecting a Field of Potatoes 71
Fig. 6.2 Wheat leaf and stem rusts by Brittlebank. From McAlpine's *Rusts of Australia* (1906), Plate A 72
Fig. 6.3 Title Page of *Zeitschrift für Pflanzenkrankheiten* Vol. 1 (1891), Listing the 44 members of *International Phytopathological Committee* 76

Fig. 7.1 Painting of Bitter pit by Margaret Senior. From Agfact H4.AC.1, Second ed. J.A.Dart 2004 82
Fig. 7.2 Bitter pitted Apples. Agfact H4.AC.1, Second ed. J.A.Dart 2004 83
Fig. 7.3 Stigmonosis in Apples. McAlpin's Second Progress Report on Bitter Pit, Plate XXXI (1913) 84

Fig. 8.1 Painting of the Codlin Moth from Charles French (1891) *Destructive Insects of Victoria* Vol. 1, Plate II 98
Fig. 8.2 Painting of the Harlequin Bug from Charles French (1891) *Destructive Insects of Victoria* Vol. 1, Plate IX 103

Fig. 10.1 Painting of bitter pitted apples by C. C. Brittlebank 115
Fig. 10.2 Vascular systems of a pear showing stone cells stippled (Fig. 78) 116
Fig. 10.3 Vascular systems of pear (figs 86 and 87) and apple (figs 88 to 94) 117

Fig. 11.1 Transverse sections showing vascular system of the apple 124
Fig. 11.2 Longitudinal sections showing vascular system of the apple 125
Fig. 11.3 Transverse and longitudinal sections of pear and quince. Main veins indicated, stippling highlighting aggregations of sclereide 128
Fig. 11.4 The McAlpines enjoying Olinda's peace with the Laidlows in 1912 130

Fig. 12.1 Median TS and LS sections of *Cleopatra* (*upper set*) and *Yates* (*lower set*) apples, dried to highlight the arrangement of the primary vascular bundles 134
Fig. 12.2 Symptoms of bitter pit in *Cleopatra* (*left*) and *Schroeder* (*right*) apples 142

Fig. 14.1 Group at the British Association for the Advancement of Science Congress, Melbourne 1914. Those in the photograph referred to in various chapters are: Back Row L to R 2nd Isabel Cookson, 3rd Ethel McLennon, 12th E.E. Pescott, 13th Daniel McAlpine, 14th C.C. Brittlebank. Centre Row 4th A.J. Ewart. Front Row 2nd C.A. Topp 154

Fig. 16.1 Harry Marshall Ward (1854–1906). G.C. Ainsworth (1981) *Introduction to the History of Plant Pathology*, Fig. 20, p 58..... 177

Fig. 18.1 Professor A.J. Ewart (1872–1937) 192
Fig. 18.2 Charles Brittlebank at work, 422 Flinders Street 202

Fig. 19.1 McAlpine's appeal to Premier Peacock 208
Fig. 19.2 McAlpine family on the steps of *Merchiston*, Toorak in 1911..... 213
Fig. 19.3 McAlpine family in front of *Invercliffe*, Wandin in 1915..... 213

List of Tables

Table 10.1 Production of apples in the commonwealth of Australia............ 114
Table 10.2 Total apple exports 1909–1911 114
Table 10.3 Accounts for the investigation for the financial year 1st July 1912 to 30th June 1913.. 121

Table 12.1 Condition of apples in the *SS Somerset* cool store consignment, on arrival in London [10th April], then after 2 weeks [25th April] and 4 weeks [8th May] held at ambient temperature 141

Table 15.1 Results from McAlpine's manorial trial (*The Argus* 11-3-1915).. 162
Table 15.2 Results from McAlpine's pruning trials (*The Argus* 20-4-1916).. 162

Table 16.1 Results of a small trial run by McAlpine (1910)......................... 170
Table 16.2 Apple production in Victoria, 1904 to 1920 – before, during, and after the investigation 171
Table 16.3 Apple production 1909–1920 172
Table 16.4 Changes in apple production and revenue generated between periods 1914/1917 and 1918/1920 172

Table 17.1 Control recommendation for bitter pit today and a century ago 183

Part I
Origin and Development of a Pioneer Plant Pathologist

Daniel McAlpine began his working life as a student teacher under his father's tutelage. Aged 20, he found a clerical job with an iron works in Cumnock, where his experience awoke an interest in geology and science generally, which in turn took him to London. There, under the influence of Thomas Huxley and William Thiselton-Dyer, his interest in biology was aroused. Through their influence he was appointed to a teaching position in Edinburgh where his talent as a lecturer soon resulted in his gaining teaching contracts in four august establishments in that city. After he married, however, he found his income was not enough to support a growing family, forcing him to seek better opportunities in Britain and abroad. Having set out for Tasmania in 1884, the family arrived in Melbourne and settled there, where he soon found a teaching position in Kew, then at Ormond College and a little later at the Victorian College of Pharmacy. During this period his interest in plant diseases was stimulated by the enormous losses farmers suffered through serious epidemics of wheat rust. As a consequence of his interest in the natural sciences, the esteem of the Master of Ormond College, the Principal of the Victorian College of Pharmacy and the impression gained by Baron Sir Ferdinand von Müller, then Victoria's Government Botanist, and the Hon. Alfred Deacon, the Victorian Minister for Agriculture, he was appointed Victoria's Consulting Vegetable Pathologist. Over the following 20 years he established himself as one of the world's leading mycologist-plant pathologists and laid the foundations of plant pathology in Australia.

Chapter 1
Introduction

1.1 Greatness Forgotten

Various tributes have been paid to Daniel McAlpine as a man (Anon. 1932), as a pioneer of Australian plant pathology (Fish 1976; White 1986), and as one of the greats among the world's plant pathologists of the late nineteenth and early twentieth centuries Willis (1949). Professor Whetzel (Fig. 1.1), who in 1907 established the first Department of Plant Pathology at Cornell University in the USA, wrote *An Outline of the History of Plant Pathology* (Whetzel 1918) in which he claimed that the only two English plant pathologists of importance in the nineteenth century were Marshall Ward and Daniel McAlpine. In his book, *Advance of the Fungi*, E.C. Large (1940) acknowledged the lead given by McAlpine to the international effort to overcome the ravages of black stem rust of wheat, which had plagued civilisation since well before Roman times. Tributes have been paid to the vast amount of high quality work that he did, which at the time was universally lauded, and which has stood the test of time in terms of his insights and the accuracy and usefulness of his work. His wisdom and sound advice were important regarding the establishment of the first Chair of Biology at the University of Melbourne (McFarlane 1924) and to Commonwealth and State agencies responsible for the investigation and control of plant diseases, which at the time were causing crop losses measured in millions of pounds (Fish 1970). These facts were either unknown beyond those involved or have faded from the collective memory. As Willis (1949) lamented, it might have been expected that at the end of his career he would have retired with a great sense of achievement and satisfaction, and been hailed as a National Treasure. Dr Jim Willis[A1] was one of Australia's leading botanists of the twentieth century (Aston 1996).

Sadly that was not the case, as McAlpine retired without as much as a thank you from his political masters. He provided 21 years of loyal service as consultant to the Victorian Department of Agriculture, while assisting most other states as well until

D.G. Parbery, *Daniel McAlpine and The Bitter Pit*,
DOI 10.1007/978-3-319-09552-3_1

Fig. 1.1 Herbert Hice Whetzel (1877–1944)

Fig. 1. Herbert Hice Whetzel (1877–1944).

1911. Then from 1911 he reluctantly put aside his love of mycology and plant pathology to investigate the cause and control of a disease caused by a physiological disorder, primarily of apples, known as bitter pit. He took the investigation as far as was then possible, but his Australian peers did not appreciate that at the time. Progress stalled in the investigation because of limited knowledge of plant physiology, especially relating to mineral nutrition. Even so he was able to provide growers and traders with advice which reduced their losses to manageable proportions. At the end of 1915, the Commonwealth Advisory Committee of Bitter Pit was somewhat reluctant to recommended extending McAlpine's contract into 1916. The Committee was critical of some of his methods and claimed his conclusions were based on unreliable data. Because of the strong support of orchardists and Departments of Agriculture in several States, however, his contract was extended for 12 months to enable him to complete aspects of the investigation. While the extension was approved, dithering over making a decision, and agreeing on McAlpine's salary, which was reduced, little of the 1915/'16 season was left in which to achieve anything.

At the end of 1916 McAlpine first claimed he had taken the investigation as far as possible and wished to return to his job as Vegetable Pathologist. According to his family, prior to undertaking the bitter pit investigation he believed he had a gentleman's agreement with the Minister for Agriculture that assured his return to the job of Vegetable Pathologist for Victoria once the investigation ended. To his great chagrin, however, that did not happen. At the end of his extended contract on 30th December 1916, and after futile attempts to either regain his old job or gain support for continuing an apple-breeding program, he returned home a disappointed and disillusioned man, telling his wife and daughters that nobody wanted him anymore (Wedge 1984).

The treatment of McAlpine was a national disgrace, not only because of his contribution to plant pathology in Australia, but also because his conclusions about the control of bitter pit were sound and greatly valued by industry and horticultural officers both in Australia and overseas.

1.2 Documents and Letters

The discovered documents, referred to in the *Preamble*, cover the first two decades of the twentieth century, and confirm McAlpine's success in the achievable half of his brief relating to the bitter pit investigation. When trolling through commentaries in his books, letters and government files housed in the State Library of Victoria, one is struck by his perceptiveness, common sense, and wide knowledge of science. The rigour and accuracy of his work and the energetic application of his knowledge, experience and perception attest to his brilliance. The reactions of his students and colleagues to his warm friendship and lecturing ability demonstrate his great ability as a scientist, teacher and communicator.

Documents in the files, however, also demonstrate that these attributes and McAlpine's public standing bestirred strong antagonism in a young, ambitious English botanist appointed jointly to the first Chair of Botany at the University of Melbourne and as Victorian Government Botanist. Within 2 years of his dual appointments in 1906, Professor Alfred Ewart's antagonism to McAlpine was evident. From 1910, it is also evident that Ewart hoped to lead the bitter pit investigation and when thwarted, presented an untested hypothesis about the cause of bitter pit, in an attempt to stall or stop McAlpine's investigation. Initially this threatened McAlpine's proposed investigation and questioned his suitability to lead it. When that failed, Ewart maintained the validity of his own hypothesis, for which no evidence was ever forthcoming, and took every opportunity to demean McAlpine's standing. By the end of 1915, although the benefits of McAlpine's investigation were recognised within the horticultural industry here and abroad, his political masters failed to appreciate the beneficial outcomes it delivered. In the political sphere, McAlpine was isolated and ignored. It is ironic, that on 29th May 1917, *The Argus* noted that the production of apples in the Commonwealth of Australia during 1916 was a record.

[**NB.** Author's Note – In order to avoid the intrusion of annoying repetitions of the positions of significant players, or explanations of bystanders' backgrounds and activities into the narrative, which is primarily the story of Daniel McAlpine's career, brief biographical notes are provided for ready reference in Appendix 1, signified in the text by superscript 'A1'

It should also be noted that as most of this book is based on previously unpublished information in the form of letters and documents, which are held in the State Library of Victoria, Melbourne's *La Trobe Australian Manuscript Collection*, unless stated otherwise. DGP.]

References

Anon. (1932) Death of Daniel McAlpine. Australas J Pharm, Vol.13 October:993, November:1003–1005
Aston HI (1996) Obituary: Dr James Hamlyn Willis AM 1910–1995. Muelleria 9:1–4
Fish S (1970) The history of plant pathology in Australia. Annu Rev Phytopathol 8:13–36
Fish S (1976) Daniel McAlpine – a pioneer plant pathologist in Australia. Aust Plant Pathol Newsl 5(1):11–14
Large EC (1940) The advance of the fungi. Jonathan Cape, London, 488 pp
McFarlane J (1924) Letter written [21-10-'24] in regard to McAlpine's influence on the establishment of the Chair of Biological Sciences at Melbourne University
Wedge E (1984) Letters [4th to 7th May-1984] to D.G.Parbery, La Trobe Collection
Whetzel HH (1918) An outline history of plant pathology. W.B. Saunders Co. Philadelphia/London, 130 pp
White NH (1986) McAlpine, Daniel (1849–1932). Aust Dict Biogr 10:193–194, Melbourne University Press
Willis JH (1949) Daniel McAlpine; in botanical pioneers in Victoria. Vic Nat 66:83–89

Chapter 2
Scotland & England

2.1 Scottish Origin and Early Years

Daniel McAlpine, was a highly intelligent and perceptive man. He learned from his experiences and applied what he learned to the benefit of those lucky enough to find themselves under his influence either in the classroom, a lecture theatre, a meeting, an excursion or through the legacy of his published work. He was a red-headed Scot, born into a typical middle class Scottish Presbyterian family, which placed great emphasis on gaining a high quality education, and the rigorous application of this to their daily lives and activities. These attitudes and values were strong among Scots from the middle of the eighteenth century, the dawn of the Scottish Enlightenment.

Daniel was born on 21st January 1849 in Dalry, Ayrshire, Scotland, the third son of Daniel and Flora McAlpine. During the next 2 years the family moved 5 miles south west to Saltcoats 2 miles from Ardeer where his father, a staunch Presbyterian and Gaelic scholar, taught at the Ardeer School. At the time, Scotland was more egalitarian than England especially regarding the importance of and access to education as a societal right. There is little doubt that McAlpine Snr held these views. In Ardeer, young Daniel and his three brothers received basic educations. Daniel was grounded in Latin and Greek, which he could speak, and learned French and German, of which in later years he claimed only a working knowledge. From the mid to late 1860s he taught at the Adeer School as a pupil teacher (Wedge 1984).

2.2 Fascination with Geology

When he was 20 years old he took a clerical job in the work's office of the Egheiton Company's Lugar Iron Works in Cumnock 25 miles south east of Saltcoats. According to the 1871 census, Daniel was then living in Auckenleck, not far from Cumnock, where his interest became strongly focussed on geology. The emergence

D.G. Parbery, *Daniel McAlpine and The Bitter Pit*,
DOI 10.1007/978-3-319-09552-3_2

of geology as a science began in Scotland through the observations of James Hutton during the second half of the 18th century.

According to Repcheck (2003), "Starting around 1750, a small group of academics, amateur scholars, government officials, clergymen and inventors, all about the same age and centred on Edinburgh, made broad and seminal contributions to Western collective knowledge within a single generation. This flowering of philosophic theory, economic processes, analyses of history and especially scientific work, is now known as *The Scottish Enlightenment*". The men involved often gathered at *Newhailes*, the grand home of Sir David Dalrymple, at Musselburgh a little out of Edinburgh, to philosophise about new-age ideas. Among the enlightened gentlemen were Adam Smith who developed the field of modern economics, Joseph Black who isolated carbon dioxide and was among the pioneers of modern chemistry, and Black's former assistant, James Watt whose perfection of the practical steam engine essentially powered the industrial revolution. Another much less known but critically important thinker was James Hutton, whom Repcheck referred to as *The Man Who Found Time*. His fascination with the nature of rock types, the contortions of rock layers and the resultant topography led him to the conclusion that the age of the earth was immeasurably greater than the 6,000 years laid down by the Church. Charles Darwin later acknowledged that it was Hutton's insight that provided the key to his developing the theory of evolution (Repcheck 2003). Hutton was an excellent speaker and could explain his ideas clearly, but he was no author which rendered his book *Theory of the Earth with Proofs and Illustrations* published in 1795, incomprehensible to most readers and it remained for Charles Lyle (Repcheck 2003) to launch the science of geology into the public arena with the publication of his book, *Principles of Geology*.

During his three plus years in Cumnock, Daniel looked into the possibility of making Geology his career and at the end of January 1872 he wrote to Professor Archibald Gieke, occupant of the foundation Chair of Geology at Edinburgh University, about the prospects of a career in geology. Geike (1872) replied that apart from the geological survey of Great Britain, which was then in progress, prospects were limited, but suggested that he might consider undertaking his own survey of an ore outcrop in a chalk deposit near Cumnock and write it up. Presumably Gieke considered that such a study if well done, might impress a mining company sufficiently to gain McAlpine entry in to the field. McAlpine decided otherwise but it seems that it was his interest in geology that inspired him to seek entry into the Royal School of Mines the following year, and it is evident that his initial choice of subjects favoured that path. At the beginning of 1879, Professor A.C. Ramsay (1879) wrote from the Royal School of Mines confirming McAlpine's obtaining a first class pass in geology. His interest in geology persisted for many years.

Daniel and his brother Archibald were similar in temperament and ability. Each became highly competent lecturers in the natural sciences and, in later life each was highly respected by their students and peers as experts in their fields. Each was esteemed for his inspirational teaching, and students in both Scotland and Australia remembered them fondly. It seems however that their primary education took different paths. While Daniel had his basic education in Ardeer, as a young boy Archibald was sent to the Free Church School in Dumbarton, north west of Glasgow.

2.3 Undergraduates in London

During Daniel's three and a half years clerking at the Iron Works in Cumnock, Archibald's and his interests in the sciences increased to the point where in January 1873 they enrolled at the Royal College of Science London, the Rev'd Paterson (1873) having supplied a testimony to their good character. In 1874, the Registrar of the University of London, W.B. Carpenter (1874), advised Daniel that he had matriculated as a student of the University of London at the June examination. In August the following year it was confirmed that he had attended the various courses and passed all his exams (Carpenter 1875b, c, d, e, f). All the subjects including organic and inorganic chemistry studied between 1873 and 1875 were listed but did not include botany. Other documents list botany and biology, which were studied between 1874 and 1876. Professor Sir Thomas Huxley[A1] delivered the lectures in biology, Sir William Thiselton-Dyer[A1] in botany, Professor A.C. Ramsay[A1] in mineralogy, Sir Archibald Geikie[A1] in geology, and Professor R. Ethridge[A1] in palaeontology (Wedge 1984; White 1986). He excelled in all these subjects and apparently was eligible to take out his degree by 11th August 1875, but curiously it was never conferred. No reason for this was given and no explanation has been found. It seems likely that Daniel's focus on a career in geology was redirected through the inspiring courses given by Huxley and Thiselton-Dyer, which switched his main interest from the physical to the biological sciences. This was possibly the reason he decided to complete both two-year courses in botany and biology by extending his undergraduate course from 3 to 4 years. It is also possible that the study of biology and botany were not available to non-matriculated students.

While Daniels' family believed he took out his degree on 11th August 1875, University of London records do not confirm it. McAlpine never cited any degree for himself in any publication, whereas on the title page of the *Biological Atlas*, published jointly with his brother Archibald (McAlpine and McAlpine 1880), the authors are given as "D. M'Alpine F.C.S., Lecturer on Biology and Botany, Edinburgh, Honourman in the Science and Arts Department, University of London, etc., and A.N. M'Alpine B.Sc. (Hons.) London, Lecturer in Botany, School of Medicine Edinburgh, Professor of Botany and Natural History, New Veterinary College Edinburgh, and Associate of the Royal College of Science, Ireland." Nevertheless, in spite of not holding a degree, Daniel McAlpine received a rigorous tertiary education, which served him well.

Three of his lecturers exerted substantial and long lasting influence on him. Professor Huxley not only encouraged and supported him in the production of Atlases on Biology (McAlpine and McAlpine 1880), Botany (McAlpine 1882, 1883a) and Zoology (McAlpine 1881a, b), but also through his own love of Australia and his hope to have returned there to a teaching post at the University of Sydney 20 years earlier, Huxley may have planted the idea of going to Australia in McAlpine's mind. It was through Professor Thiselton-Dyer that he met Sir Joseph Hooker who supported McAlpine' establishment in Australia, and also Mordacai Cooke who probably encouraged McAlpine to send him collections of Australian fungi and who later provided McAlpine with specimens of Australian fungi collected by early European explorers and other collectors. McAlpine maintained his friendship with Thiselton-Dyer (1904) over 30 years. Similarly Professor Gieke and McAlpine began an association in 1872, which spanned 50 years.

2.4 Teaching Atlases and English Texts

It is evident from McAlpine's comments in his Forward to the *Biological Atlas*, that during his undergraduate course he had noted a need for pictorial guides to assist students in their studies of practical biological subjects, which at the time were unavailable. It is also evident from his comments that Huxley encouraged and assisted him in providing them such that many of the macroscopic and microscopic illustrations in the *Biological Atlas* were prepared from specimens dissected and studied in the Biological Laboratory of the Royal School of Mines, London. These illustrations were in watercolour, carefully labelled, and explained in English!

In the introduction to the 1989 reproduction by Bracken Books, London, *of The Botanical Atlas – Guide to the Practical Study of Plants* by Daniel McAlpine, Dr. S.M. Walters (1989) described the intellectual environment of the era during which McAlpine found his way into botany. He also outlined the development of botanical science during the nineteenth century and in particular the fascination with life revealed under the microscopic. He described the excitement aroused through the study of cryptogamic botany, the manner in which it was taught and the intellectual attraction it engendered. Walters made the point that, while the morphology and anatomy of flowers and fruits of Phanerogams were generally well know by the sixteenth and seventeenth centuries, comparative knowledge of the Cryptogams was limited prior to about 1840, after which a revolutionary advance in interest and knowledge occurred. The widespread epidemic of the potatoes murrain (Large 1940) throughout Europe and the discovery of its cause (Berkeley 1846) no doubt played important roles in the strong rise in interest in microorganisms. But so too did developments in Europe.

Walters (1989) linked this burst of interest and enterprise, to the growth of German scholarship that accompanied the development of scientifically and technologically minded Universities, which gradually replaced the older classical ones. These Universities sprang up and flourished between the end of the Napoleonic Wars after 1815 and the unification of the German States in 1871. By then, the development of the microscope had reached a point, which allowed detailed study of cryptogamic organisms, with much of the resultant literature published in German. In 1839, Prince Albert of Saxe-Coburg and Gotha married Queen Victoria of Britain. Prince Albert had a keen interested in and a broad knowledge of the emerging sciences and industrial developments in Europe and became a driving force in the development of British science and industry. His activities created such interest and awareness that British students sought to complete their scientific education by attending one of the great German Universities of the day under such

Professors as de Bary at Strasburg, von Nägli at Münich, Pringsheim in Berlin, Sachs in Würtzburg and Strasburger in Jena, the city where Carl Zeiss founded his optical equipment factory in 1846, and produced arguably the most advance microscopes of the time. However, literature and texts produced from these institutions were all in German and thereby not accessible to most British students.

By the early 1870s, Thomas Huxley and Thiselton-Dyer were teaching biology and botany respectively to a new generation of biologists destined to teach a new science curriculum throughout the newly established National Schools introduced after the Education Act of 1872 (Walters 1989). That same year Huxley transferred his Department of Science, originally called the Royal School of Mines and Science, to South Kensington (Walters 1989) but must have retained his laboratories at the Royal School of Mines. At the time, however, the best modern texts were available only in German, as was the case when Daniel McAlpine became interested in biology and found there were few modern text books in English and no texts at all for teaching practical courses in biological subjects. Thus from about 1876, Daniel and Archibald McAlpine began work on providing such help.

In 1878, Dr. Arnold Dodel-Port at the University of Zurich and his wife Carolyn began publishing a series of first class texts and illustrations for the production of wall charts used in lecture halls and as illustrations for a *Botanical Atlas* (Dodel-Port and Dodel-Port 1878–1883), which was published in ten sections over 5 years. It is now evident that prior to McAlpine's completing his studies at South Kensington, either on his own initiative or at the suggestion of Huxley, his brother Archibald McAlpine and he began preparing their own *Biological Atlas* plus wall charts to provide illustrations to compliment the lectures of Huxley and others. In the acknowledgements made in their *Biological Atlas* (McAlpine and McAlpine 1880), the brothers stated that their illustrations were prepared either from dissections and microscopic preparations made in the Biology Laboratory at the Royal School of Mines or copied from reliable sources, all of which were acknowledged at the top of pages containing the legends to plates and figures in the atlas. Preparation of their *Biological Atlas* appears to have begun before McAlpine became aware of what the Dodel-Ports were doing because 3 years later, in the preface to Volume 2 (Cryptogams) of the *Botanical Atlas*, dated December 1882, McAlpine (1883a) expressly thanked Professor Dodel-Port for allowing him free and full use of the figures in his *Anatomical and Physiological Atlas of Botany* (Dodel-Port and Dodel-Port 1878–1883). Although the Dodel-Ports would have begun preparing material for their atlas at much the same time as the McAlpine brothers began preparing theirs, there is no acknowledgements of the Dodel-Ports in the McAlpines' *Biological Atlas* suggesting it was begun on the McAlpines' own initiative or following a suggestion by Huxley.

At the top of the legends to the illustrations in the two volumes of the *Botanical Atlas*, McAlpine (1883a) acknowledged the source of each illustration. For some, however, he made no acknowledgement indicating that he was the artist (Figs. 2.1 and 2.2). For others, he states that the illustrations were adapted from those of others often two or more authors (Figs. 4.1 and 5.5), while some were exact copies (Fig. 2.3).

After McAlpine began teaching in Edinburgh in 1875, his awareness of the need to have up-to-date modern supporting literature, available in English, resulted in his

Fig. 2.1 Frontice Piece from McAlpine's Botanical Atlas Vol. 1 (1882) (Courtesy of the Royal Botanic Gardens Library, Melbourne)

writing *Short Notes for Biology Students* (McAlpine 1877). Then, having discovered the Dodel-Ports' publications, and noting that the reviewer for *Nature*, had acclaimed the Dodel-Ports' Atlas the finest botanical text of its day, and saying it should be available in English, McAlpine began translating it. Dodel-Port's *Erläuternder Text aus Anatomisch-physiologischen Atlas der Botanik für Hoch- und Mittleschulen* was a 424 page document, and while he translated it along with legends and labels to figures McAlpine compiled four Atlases of his own (McAlpine 1881a, b, 1882, 1883a) comprised 240 pages of text and 70 Plates containing 1,388 coloured Figures, many produced by McAlpine himself, all accurately labelled and accompanied with brief explanatory notes. Finally he wrote *Atlas of Elementary Physiology*

Fig. 2.2 The Sweet Pea Botanical Atlas Vol 1, Plate XXI (1882) (Courtesy of the Royal Botanic Gardens Library, Melbourne)

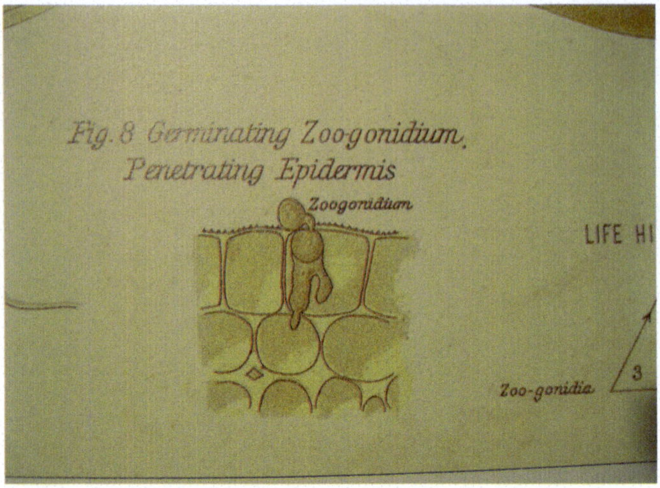

Fig. 2.3 I infection hypha of *Phytophthora infestans* copied from De Bary. Botanical Atlas Vol. 2, Plate XII, Fig. 8 (1883) (Courtesy of the Royal Botanic Gardens Library, Melbourne)

and Physiological Anatomy (McAlpine 1883b) basically a rewrite of the Dodel-Ports' work, but with his own amendments and additions.

Walters (1989) lauded the importance of McAlpine's atlases, particularly the two part *Botanical Atlas* claiming that by 1883, when the *Botanical Atlases* first appeared, the revolutionary advance in knowledge of and interest in cryptogamic botany had happened and an exciting new evolutionary picture of the whole Plant Kingdom and its essential unity of basic structure was presented, for the first time, to British students, in English. Thus McAlpine had a done a great service to all British students in the late nineteenth century.

2.5 Academia in Edinburgh

Daniel was appointed Lecturer in Agriculture, Botany, Biology and Geology at the Watt Institute and School of Arts, Edinburgh, in 1875. Ms Helen Taylor (2012), archivist for Heriot Watt University confirmed McAlpine's starting date, a year earlier than cited by other authors (Fish 1976; Wedge 1984; White 1986). One possible explanation for confusion over the date was that his initial lecture commitment was not a heavy one, and at the time he was single. Thus, provided he fulfilled his lecture commitments and because he was preparing teaching material, it appears that the Director of the Watt Institute enabled him to spent time in London completing his studies of biology and botany and working on the *Biological Atlas*.

In addition to working on the *Biological Atlas* with his brother in London, after his appointment to the Watt Institute, he attended a special course in agricultural chemistry for the instruction of science teachers held at Rothamstead Experimental

Station, and passed the examination (Anon. 1881). Rothamstead was part of the University of London, the institution which, conferred degrees and diplomas obtained through study there, as well as at other associated establishments. Because one of his duties at the Watt Institute was teaching Agriculture where knowledge of applied chemistry was of increasing importance, it is understandable that McAlpine attend the course at Rothamstead. However, he was also an able student of chemistry having gained the distinction of Honourman in both organic and inorganic chemistry during his undergraduate studies in London. Possibly as a consequence of his scholarship as a chemist, on 5th December 1878 he was admitted as a Fellow into the British Chemical Society [FCS] (Anon. 1879; Geike 1879), now a component of the Royal Chemistry Society.

McAlpine was never idle and must have left those around him breathless! From 1875 until he left Scotland for Australia he was continuously busy preparing and delivering lectures and practical classes and developing teaching charts and other texts to support his own lectures and practical classes. He augmented his lecturing income from the sale of his charts and atlases, the latter being sold cheaply to students and others. This was necessary because Isabella and he were married in 1878, established their own home in Edinburgh and begun a family, and the income from the Watt Institute alone was not enough. According to Wedge (1986) this was possibly a reason he did not take out his degree because of the costs associated with doing so. At the same time it is not known just how long after completing their studies in London that the brothers continued returning to complete the illustrations for their *Biological Atlas*. In their preface to it they state that," the Atlas was produced in recognition of the need of students to handle, dissect and study field specimens for themselves. The Atlas is provided as a guide to be use in association with texts by Huxley, Nicholson, Macalister, Sachs, M'Nab, etc." It is of some interest to note that in a letter thanking McAlpine for sending him a copy of the Atlas, Charles Darwin (1881) issued a warning that lecturers would need to be on their guard to avoid students simply copying illustrations from the atlas rather than drawing the specimens put before them!

References

Anon. (1879) Secretary, British Chemical Society to McAlpine, letter 5-12-1879 advising he had been elected Fellow of the Chemical Society; the admission fee being £4 and the annual subscription £2

Anon. (1881) Letter [30-9-1881], on Science Form 1134, headed, *Science & Arts Department of the Committee of the Council of Education, South Kensington*, listing the names of those who attended a Special course of Instruction in Agricultural Chemistry for Teachers at Rothamstead and passed Subject XXIV – *Principles of Agriculture*

Berkeley MJ (1846) Observations botanical and physiological on the potato murrain. J Hortic Soc I:9–34

Carpenter WB (1874) Letter [29-7-1874] from W.B. Carpenter, Registrar University of London, to McAlpine informing him that he matriculated as a student in the University at the June Examination in the year 1874

Carpenter WB (1875a) Letter [11-8-1874] *idem*, certifying that he had attended the course of lectures in Applied Mechanics during the session 1874 to 1875 and had passed he examinations
Carpenter WB (1875b) Letter [11-8-1874] *idem*, certifying that he had attended the course of lectures in Chemistry during the session 1873 to 1874 and had passed he examination. Signed by W. Manly
Carpenter WB (1875c) Letter [11-8-1874] *idem*, certifying that he had attended the course of lectures in Chemistry during the session 1873 to 1874 and had passed he examination. Signed by E. Frankland
Carpenter WB (1875d) Letter [11-8-1874] *idem*, certifying that he had attended the course of lectures in Geology during the session 1873 to 1874 and had passed the examination in the highest degree satisfactory. Signed by Andrew Ramsay LLD, FRS Lecturer in Geology
Carpenter WB (1875e) Letter [11-8-1874], *idem*, certifying that Daniel McAlpine had attended the course of lectures in Mineralogy at the Royal School of Mines during the session 1874 to 1875 and had passed he examination. Signed by Warrington W. Smith, Lecturer in Mineralogy
Carpenter WB (1875f) Letter [11-8-1874], *idem*, certifying that Daniel McAlpine had attended the course of lectures in Physics at the Royal School of Mines during the session 1873 to 1874 and had passed the examination. Signed by Fredrick Guthrie
Darwin C (1881) Letter written to McAlpine [10-5-1881] commenting on the Biological Atlas
Dodel-Port A, Dodel-Port C (1878–1883) Erläuternder Text aus Anatomisch-physiologischen Atlas der Botanik für Hoch- und Mittleschulen. University of Zurich, Zurich
Fish S (1976) Daniel McAlpine – a pioneer plant pathologist in Australia. Austr Plant Pathol Newsl 5(1):11–14
Geike, Prof. Sir Archibald (1872) Letter [29-1-1872] written to McAlpine, responding to his enquiry about opportunities for Geologist in the UK
Geike, Prof. Sir Archibeld (1879) Letter written to McAlpine (2-1-'79) telling him of his admittance as a Fellow of the British Chemical Society
Geike, Prof. Sir Archibald (1882) Correspondence with McAlpine [15-5-1882] with a copy of his supporting reference for the Chair of Natural History at the University of Otago, New Zealand
Large EC (1940) The advance of the fungi. Jonathan Cape, London, 488 pp
McAlpine D (1877) Short notes for biology students. Oliver and Boyol, Tweedle Court Edinburg, Glasgow and Dublin. 44 pp
McAlpine D (1881a) Zoological atlas, vol I, Vertibrata, 48 pp, 231 coloured illustrations. W. and A.K. Johnston, Edinburgh/London
McAlpine D (1881b) Zoological atlas, vol II, Invertibrata, 32 pp, 246 coloured illustrations. *ibid*
McAlpine D (1882) Botanical atlas, vol I, Phanerogams, 55 pp, many coloured illustrations. *ibid*
McAlpine D (1883a) Botanical atlas, vol II, Cryptogams, 55 pp, many coloured illustrations. *ibid*
McAlpine D (1883b) Atlas of elementary physiology and physiological anatomy. *Ibid*
McAlpine AN, McAlpine D (1880) Biological atlas, 1st edn. *ibid*
Paterson, Rev. AS (1873) This letter [9-10-1873] testified to the worthiness and good character of both Daniel and Archibald McAlpine
Ramsey AC (1879) A personal letter [30-1-1879] from Ramsey telling McAlpine his pass in the Geology exam was first class
Repcheck J (2003) The man who found time – James Hutton and the discovery of the Earth's antiquity. Simon & Schuster, UK Ltd., Gray's Inn Road, London, 247 pp. ISBN 0-7432-3189-9
Taylor HE (2012) Email to DGP including the Minute of the Watt Institute and School of Arts Board Meeting at which Daniel McAlpine's appointment was recorded
Thisleton-Dyer WT (1904) A letter [15-12-'04] written to McAlpine
Walters SM (1989) Introduction to the Bracken Books Republication of *The Botanical Atlas, a Guide to the Practical Study of Plants* (Part 1 and 2 Combined) by Daniel McAlpine, Best Seller Publications Ltd., Princess House, 50 Eastcastle Street, London, 136 pp. ISBN 1 85170 255 5
Wedge E (1984) Letters [4th to 7th May-1984] to D.G.Parbery, La Trobe Collection
White NH (1986) McAlpine, Daniel (1849–1932). Austr Dict Biogr 10:193–194, Melbourne University Press

Chapter 3
Highly Educated Non-graduate

According to family records (Wedge 1984), McAlpine was appointed Professor of Natural History in the University of Edinburgh from 1876 until 1884. They also claimed that he lectured in natural history at the new Veterinary College in Edinburgh from 1876, in biology and botany at the Edinburgh Watt Institute from 1877, and that he gave lectures in biology in the Pharmaceutical Department of the Edinburgh School of Medicine (White 1986). While there was confusion over the dates, and the probable courtesy use of the title "Professor", it is evident that the substance of these articles is correct. They also believed that he had been offered an honorary Doctor of Law degree in 1884, which he declined on the grounds that this high qualification in such a land of social equality as Australia could be disadvantageous. Wedge (1984) said he greatly regretted his decision in later years.

A request for information from the University of Edinburgh regarding the alleged offer of a LLD *honoris causa tantum* resulted in the astonishing response (Eddie 2011) that Daniel McAlpine was never an employee of the University of Edinburgh. The Professors of Natural History, during the years when Australian records claim McAlpine held the post, were Professor Sir Wyville Thomson[A1], well versed in botany geology physics and zoology, from 1870 to 1882, and Professor J. Cossar Ewart[A1], a zoologist, was Professor of Natural History from 1882 until 1927. McAlpine's name was not listed among assistants to the Professors, nor was there any record of his lecturing in natural history at the New Veterinary College which opened in 1873, nor of his lecturing in the Pharmacy Department of the Edinburgh School of Medicine. Nor was there any record of his being offered an honorary doctor of laws. All of these assertions have been proved false apart from the offer of an Honorary Degree, which is still open to question.

D.G. Parbery, *Daniel McAlpine and The Bitter Pit*,
DOI 10.1007/978-3-319-09552-3_3

3.1 Proof of Appointments in Edinburgh

Proof of all his appointments in Edinburgh was found in McAlpine's correspondence held in the LaTrobe Collection of the State Library of Victoria, whereas there was no information on McAlpine in the Edinburgh University archive. On 5th May 1881, Sir Wyville Thomson wrote a letter supporting McAlpine's application for the position of lecturer in biology at the University of Nottingham (Thomson 1881). Thus McAlpine was known to Sir Wyville as a lecturer and is likely to have given lectures at his request. Further evidence of McAlpine's association with the University is contained in a letter of 2nd January 1885, in which R. Urquart (1885), Lecturer in Materia Medica and Secretary of the Pharmacy Department of the School of Medicine, Edinburgh, wrote in support of McAlpine's application for a lectureship at the Melbourne School of Pharmacy, stating: " Mr Daniel McAlpine lectured in Botany and Natural History in this Institution for some time previous to his leaving the country. He is sure to give every satisfaction". Finally in an article on the history of the New Veterinary College in Edinburgh, Warwick and Macdonald (2003) state that Daniel McAlpine taught Natural History and Botany at the College from 1879 until 1984 when he resigned. His brother Archibald was appointed in his place for the following 6 years. Archibald had been previously appointed there soon after his graduation from about 1875 to 1978.

In view of the fact that all the claims made by McAlpine's family about Daniel's activities and appointments in Edinburgh have been confirmed, their belief in his being offered an honorary degree has to be left open. It is quite possible that he was offered an honorary degree, perhaps a D.Sc. and because it was the University's tercentenary, commemorated in part by the award of several LLD *honoris causa tantum* degrees, the one they believed he declined was confused by his family with those in the commemorative list. Had Sir Wyville made the offer, his sudden death in 1882 could explain why there is no record of it. Similarly had Professor J.C. Ewart simply made a verbal offer and McAlpine declined it, it is unlikely that there would have been any formal record of the offer been made.

In February 2012, evidence of McAlpine's appointment as lecturer in Natural History at Heriot-Watt University was confirmed by the Archivist, Ms Helen Taylor (2012) who wrote: "I have checked the records for the Edinburgh Watt Institute and School of Arts, which was the precursor to Heriot-Watt College [now University]. Mr. D. McAlpine lectured between 1875 and 1884 in Botany, Geology, Agriculture and Biology, and the Annual Reports note that Mr. Daniel McAlpine, Glasgow, was appointed to teach Botany and Geology in 1875. The reports also noted his resignation in 1884". His appointment was styled, "Lecturer in Natural Science". McAlpine (1877) referred to his employment at Watt College in his first publication and several following ones.

3.2 Academic Career

McAlpine's (1912) listed of 221 publications between 1877 and 1912, which was outstanding. In a note of introduction to his list of publications, he stated "Ever since I studied biology under the late Professor Huxley, and botany under Sir William Thiselton-Dyer at the Royal College of Science, London, these subjects have engaged my attention, and from the *Short Notes for Biological Students*, published for the use of students attending my lectures at Heriot-Watt College, Edinburgh, in 1877, down to my last work in 1910 on the *Smuts of Australia*, a wide range of subjects and a great variety of practical problems have been dealt with". It seems that although he was never a member of staff of other institutions in Edinburgh, he was engaged to provide lectures and practical classes on a contractual basis, as indeed was most of his work in Australia from 1884 until 1916. This could account for his not being recorded as a staff member of any Scottish establishment other than Heriot-Watt College. In 1877 he was living in Park Terrace, Edinburgh, where he received the account for the publication of 350 copies of *Short Notes for Biological Students*. The printing cost £7/5/-, plus alterations to proofs of £1/5/- extra, payable to Lorimer and Gillies – Steam-Power Printers, Lithographers and Engravers, Edinburgh.

His commitment to providing top quality lectures is demonstrated in a letter written to McAlpine on 5th January 1883 by Littleton Wilday Esq. of *L. Wilday & Company, Lithographers & Printers* of 9 Castle Street, Holborn, London. It illustrates how McAlpine took great trouble to provide well-produced, highly visible wall charts on aspects of natural history for use during his lectures, and for sale to other teaching establishments. McAlpine had requested Wilday to produce large-scale charts of the human body. Wilday (1883) wrote back to advise that although they could produce charts 7 ft and 6 in. by 5 ft [2.3 × 1.54 m], for which evidently McAlpine had asked, they recommended charts of 5 ft 3 in., by 3 ft 10 in. [1.62 × 1.2 m] because the smaller charts were more portable. In view of the fact that he lectured at various establishments in Edinburgh, portability would have been an important consideration. Some if not most of these charts were adjuncts to the Atlases and they were visible from the back of lecture halls. He probably used his charts when lecturing in Melbourne, and in his first report as Vegetable Pathologist (McAlpine 1890) he use the chart of the rust life history to educated farmers about the biology of rust fungi.

It is evident that from the beginning of his career McAlpine focused on high quality outcomes. Initially he demonstrated this through his commitment to delivering high quality lectures and in producing practical teaching aids for students. *Short Notes for Biological Students* of 1877 and his series of meticulously prepared atlases published between 1880 and 1886 were highly prized. They were produced to enhance understanding of plants and animals during detailed studies of them. His desire to provide students with the best possible assistance in gaining knowledge of and understanding the natural world in all its forms, is expressed in the preface to the first edition of the *Biological Atlas* where McAlpine (1880) states:

> No apology is needed for issuing a work of this kind. While there are large and well-executed wall-charts of this kind which the student may admire from a distance, there is no set of drawings at a convenient price and in a handy form which can be used as a map to show him the various regions of the plant or animal, and guide him in his attempt to explore them. It is hoped that the cheap and convenient form of the Atlas may favour its introduction even into higher-class schools. There the study of biology might be made a powerful instrument, both of information and training, and a valuable preliminary to the study of physiology.

His small textbook in English, a rarity at the time, and the evident care he took in preparing lectures and his confident energetic delivery combined with wit and humour attracted students to his subjects and to him. Over the years students showed their appreciation in various ways. For example, on 19th April 1878, possibly at the time of his marriage, his students at the Watt Institute and School of Arts showed their appreciation by presenting him with a purse of sovereigns, a very rare tribute indeed.

The Atlases proved popular and were widely used. A second edition of the *Biological Atlas* was produced in 1881. Dr. S.M. Walters (1989) was closely acquainted with the quality and value of McAlpine's books and wall charts. He joined the staff of the Botany School, University of Cambridge as Curator of the Herbarium in 1949, lectured in botany from 1962 until 1973 when he took on the Directorship of the University Botanic Gardens, Cambridge, continuing to give some lectures until he retired in 1983. Throughout his 20 years of lecturing, he continued to find McAlpine's wall charts of value. In his tribute to McAlpine he said that although the idea of the *Botanical Atlas* was derived from the work of Professor Dodel-Port, a comparison of McAlpine's (1883a) book with Dodel-Ports' original, showed that much of the *Botanical Atlas* represented the talents of a brilliant and inspired young lecturer.

Although correspondence relating to the publication of each of his teaching atlases has not been found, they were all published by W. & A.K. Johnston, Geographers and Engravers to the Queen for Scotland, Publishers of Geographical and Educational Works, Edinburgh between 1880 and 1886. It is probable that they were all published under the same terms as Daniel's two-volume *Botanical Atlas*, Phanerogams in 1882 and Cryptogams in 1883. A letter from Johnstons (1881) set out the terms for payment which was a total of £150 to be paid in three instalments, each of £50, the first after the first volume was submitted, the second after the second volume was submitted, and the final one following publication.

In early 1881 McAlpine began negotiations with W. Swan Sonnenschein & Co. (1883) to publish *Life Histories of Plants*, a textbook rather than an atlas. W. Swan (1883) decline to pay a lump sum on the grounds that such texts required regular updating, thus they offered him 10 % of sales. The tyranny of distance possibly made this, the slowest of his books to appear in print due to difficulties in obtaining blocks. These were required to provide some of the illustrations, and were borrowed from authors who had used them in their own publication. The time required for proofs to reach McAlpine from Britain and again on their returned from Australia would have been at least 3–4 months. In the event, when it became available in

1886, it was hailed as a very valuable teaching aid, especially by pharmacy students. One copy is held in the library of the Victorian College of Pharmacy and another in the Latrobe collection.

McAlpine promoted his abilities as a lecturer and his knowledge of biology generally by distributing copies of his books to biologists of note and those in positions of authority who might have helped further his career, such as Charles Darwin (1881). Most were complementary and saw his books as valuable, authoritative teaching aids and praised their quality. He continued this practice throughout his career. In addition, he dedicated each atlas to those responsible for his employment. For example he dedicated the *Botanical Atlas* to the Right Honourable the Earl of Rosebury LLD. FRSE., etc., Lord Rector of Edinburgh University. In a similar vein he dedicated his *Zoological Atla*s to The Honourable Lord Shand, President of the Watt Institute and School of Arts. In each case he had sought and obtained their permission to do so. It was through his obtaining letters of introduction from such highly placed contacts to the Governor of Tasmania (Donaldson 1884) and the Curator of the Botanical Gardens Hobart (Hooker 1884) before leaving England that he obtained the support of the Premier of Victoria for his appointment to the lectureship at Ormond College in 1885.

3.3 Search for Greener Fields

While there is ample evidence that his employers in Edinburgh recognised his scholarship and lecturing ability, they were frustrated by not being able to pay him enough to retain his services. Thus testimonies to the quality of McAlpine's teaching were generally expressed in letters supporting his applications for teaching posts elsewhere. On 27th January 1880, Professor James Donaldson, Director of the Watt Institute wrote in support of his application for the Chair of Natural History at the University of Ottawa in Canada (Donaldson 1880), stating that his lectures were clear, knowledgeable and up-to-date, and demonstrated his intense love of his subject. Two years later, when McAlpine's students heard that he had applied for the Chair of Biology and Curator of the Museum at the University of Otago in New Zealand, 30 of them signed a statement in support of his application extolling his merit as a lecturer, and among other things, the fact that he was extremely widely and well read (Anon. 1882). Sir Archibald Gieke also supported his application for the Chair in Otago, saying "that Daniel McAlpine will not merely instruct his students but will infuse them with enthusiasm for his subjects" (Geike 1882). Further evidence of his teaching abilities is found among tributes published in the *Australian Journal of Pharmacology* (Anon. 1932) following his death. Several staff members and past students paid glowing, affectionate tributes to a kindly, energetic, engaging man whose long-lasting impact they highly valued. These are dealt with later (Chap. 4.3).

In spite of his love of teaching and the satisfaction he derived from it, the financial side of his life in Scotland was difficult. In a reference written to help McAlpine establish himself in Melbourne, Rev. John Glasse (1884), Minister of Old Greyfriar's Church, Edinburgh, and a member of the Board of Directors of the Watt Institute referred to the enjoyment of McAlpine's lectures by his students and their respect for him. Also that he personally was sorry that the College had been unable to pay more to retain this man of unblemished character. In another letter of introduction, this time to the Governor of Tasmania, written on 23rd May 1884, Professor Donaldson (1884) said he wished he could have been able to pay more to retain McAlpine's services.

After he had reached Australia, McAlpine received a letter from the R.T. Scott (1884a), Secretary of the Watt Institute, containing the following from the Directors' meeting after his resignation: "A letter was read from Mr. McAlpine resigning (on account of his leaving for Tasmania) his appointment as lecturer in Botany, Geology and Agriculture. The Director accepted with regret Mr. McAlpine's resignation, and on the Motion of Mr. MacBurn unanimously agreed that a gratuity of fifteen guineas [£15/15/00=$31.50, ca $2,000 today] be given to Mr. McAlpine in consideration of the small remuneration he has had while lecturer at the Institute."

McAlpine had formally written to the Board of Directors a month or more before his departure, telling them of his intention to emigrate, and proposing the position might be offered to his brother Archibald. On 3rd December Scott (1884b) wrote again to say that the Board of Directors had agreed to his proposal and his brother would continue the classes after he had left till the end of the summer session. He added that the Directors regretted their inability to pay McAlpine more.

References

Anon. (1882) Letter [undated] signed by 30 students from Watt Institute and School of Arts supporting McAlpine's application for the post in Otago, New Zealand

Anon. (1932) Death of Daniel McAlpine. Austr J Pharm, October:99, November:1003–1005

Darwin C (1881) Letter [10-5-'81] to McAlpine commenting on the Biological Atlas

Donaldson, Prof. J (1880) Letter [27-1-1880] written by the Director, Watt Institute and School of Arts, supporting McAlpine's application for the Chair of Natural History at the University of Ottawa, Canada

Donaldson, Prof. J (1884) Letter [23-5-'84] written to the Governor of Tasmania, with a testimonial to McAlpine's ability as a natural science lecturer

Eddie GD (2011) Email [19-9-'11] sent to DGP from Dr. Eddie, Assistant Librarian Archives and Manuscripts, Special Collections, University of Edinburgh Library, regarding lack of information regarding D. McAlpine

Fish S (1976) Daniel McAlpine – a pioneer plant pathologist in Australia. Austr Plant Pathol Newsl 5(1):11–14

Geike, Prof. Sir Archibald (1882) Correspondence with McAlpine [15-5-1882] with a copy of his supporting reference for the Chair of Natural History at the University of Otago, New Zealand

Glasse J (1884) Testimonial [3-12-1884] supporting McAlpine in seeking appointments in Melbourne

Hooker, Sir Joseph (1884) Letter [26-5-'80] written to Mr. Abbott, curator of the Hobart Botanical Garden, with a testimonial to McAlpine's abilities

McAlpine D (1877) Short notes for biological students. Oliver and Boyd/Tweedale Court, Edinburgh/Glasgow/Dublin, 40 pp

McAlpine D (1880: with A.N. McAlpine B.Sc) Biological atlas, 1st edn. W. and A.K. Johnston, Edinburgh/London

McAlpine D (1890) Report of the vegetable pathologist. Report of the Department of Agriculture, Victoria. Government Printer, Melbourne

McAlpine D (1912) List of scientific works, papers, bulletins etc. 1877–1910. Government Printer, Melbourne

Scott RT (1884a) Mr Scott, Secretary, Watt Institute and School of Arts wrote [15-4-1884] to McAlpine expressing regret over his resignation to go to Tasmania, telling him the Board had voted him a 15 Guinea gratuity and of his brother's appointment in his place

Scott RT (1884b) Mr Scott, Secretary, Watt Institute and School of Arts wrote [3-12-1884], sending a copy of the Board's minute of appreciation for his service, regret at McAlpine's departure, and for not having been able to afford more generous pay

Swann Sonnenscheim (1883) Letter setting out proposed payment for Botanical Atlases

Taylor H (2012) Letter [20-2-2012] sent to the author noting Daniel McAlpine's dates of appointment to and resignation from lecturing duties at Watt Institute and School of Arts, and the subjects he taught

Thomson JA (1884) Letter written [9-9-1884] to McAlpine setting out the terms for teaching at Kew High School

Thomson, Prof. Sir Thomas Wyville (1881) Letter [5-5-1881] in support of McAlpine's application for the lectureship in biology at the University of Nottingham

Urquart R (1885) Testimonial [2-1-1885] supporting McAlpine's job hunting. Urquart was Secretary of the Pharmacy Section of the School of Medicine, Marshall Street, Nicolson Square, Edinburgh, and Lecturer in Materia Medica and wrote: "Mr. Daniel McAlpine lectured in Botany and Natural History in this Institute for some time previous to his leaving the country. He is sure to give every satisfaction"

Walters SM (1989) Introduction to the Bracken Books Republication of *The Botanical Atlas, a Guide to the Practical Study of Plants* (Part 1 and 2 Combined) by Daniel McAlpine, Best Seller Publications Ltd., Princess House, 50 Eastcastle Street, London, 136 pp. ISBN 1 85170 255 5

Warwick CM, Macdonald AA (2003) The New Veterinary College, Edinburgh, 1873 to 1904. Vet Rec 153:380–386

Wedge E (1984) Letters [4th to 7th May-1984] to D.G. Parbery, La Trobe Collection

White NH (1986) McAlpine, Daniel (1849–1932). Aust Dict Biogr 10:193–194, Melbourne University Press

Wilday L (1883) Lithographers who advocated [5-1-'83] the smaller of two possible chart sizes for a full scale drawing of the section of a baby

Chapter 4
Australia Felix

4.1 Establishment in Australia

McAlpine began moves to leave Edinburgh in 1880. The economy was weak and his income from lectures and the sales of books, etc was inadequate. Between 1880 and 1882, he had applied for posts in Ottawa, Nottingham and New Zealand without success. Three of his best friends had emigrated to Australia and were urging him to follow their example. John Darrock, who later became a Presbyterian Minister based in Melbourne was one. William Gillies was another, who was a writer interested in science, and lived in Sydney. Dr. John Walker-Smith, a medical practitioner, also lived in Sydney. In 1884, following the death of their infant son John, however, the McAlpines decided to migrate to Tasmania (Wedge 1984).

Letters in the Latrobe Australian manuscript collection at the State Library in Melbourne, confirm this. Professor James Donaldson (1884), provided McAlpine with an introduction to the Governor of Tasmania while Sir Joseph Hooker (1884), Director of Kew Gardens directed one to Mr. Francis Abbott, Curator of the Botanic Gardens Hobart, commending McAlpine to him since McAlpine was intending to settle in Tasmania. It is probable that McAlpine collected Donaldson's letter before he travelled to London and collected Hooker's letter between 23rd and 26th May, prior to embarking for Australia.

On 28th May 1884 McAlpine with his wife Isabella and daughters, Claudine and Constance aged approximately 5 and 3 years boarded the steam ship *Gulf of Carpentaria* at East India Dock London, bound for Hobart, Melbourne and Sydney. More passengers boarded at Gravesend on the following morning and again at Plymouth before heading for Teneriffe on 30th May. After re-coaling in Teneriffe and delay by gales they arrived in Cape Town on 27th June where they again re-coaled and took on five more passengers. They sighted the Coast of Tasmania near Port Davey at 9 am of 22nd July, rounded South West Cape at 9 pm and arrive in Hobart early the following morning. Whether or not McAlpine visited his two contacts in Hobart is unknown. The family, however, travelled on to Melbourne,

D.G. Parbery, *Daniel McAlpine and The Bitter Pit*,
DOI 10.1007/978-3-319-09552-3_4

where they disembarked 4 days later on 27th July after a voyage of 60 days covering 12,100 miles at an average speed of 8.7 knots.

SS *Gulf of Carpentaria* was a modern ship driven solely by steam engines. Launched in 1881, it was 2,400 tons, carried up to 600 tons of cargo and over 300 passengers. The *Gippsland Times* (1884) reported that among the passengers were some children with measles, which was often fatal, so that the children were hospitalised in Hobart. The McAlpines were not among them and were permitted to travel on to Melbourne. The hazards of early travel, however, were emphasised by the wreck of the SS *Gulf of Carpentaria* 7 months later. When returning from Brisbane to England via Adelaide, on 16th September 1885 the vessel struck an uncharted rock between Cleft Island and the Anser Island group just after rounding Wilson's Promontory, the southern most tip of mainland Australia. This was reported in the *Gippsland Times* (1885) the following day. All the crew and the few passengers abandoned ship, were packed into three lifeboats and arrived safely on the shore at Wilson's Promontory a few hours later.

The McAlpine family arrived in Victoria when the State's economy was booming, the women's suffragette movement was born, large land holdings were about to be divided further to provide farms for more families establishing dairy farms and orchards. Extensive holding had already been reduced after the 1860 Robinson Land Act. McKay's *Sunshine* combine harvester had just come on the scene, and in December 1884, Melbourne was provided with coal gas for cooking and heating, and with Edison's electric light for illuminating homes and brightening up the city.

McAlpine lost little time seeking employment. On 9th September J.H. Thompson (1884) wrote, arranging for him to provide sessions of instruction on the flora of Australia, especially Victorian flora, for the enlightenment of students at Kew High School. This might have seemed a daunting task for someone who had hardly had time to explore Victoria's flora, but he had attended lectures by Professor Thomas Huxley who held an abiding interest in Australia, and knowing Professor Thiselton-Dyer, and Sir Joseph Hooker well enough to ask him for a testimonial, it is likely that he had at least a passing knowledge of some Australian genera as well as his fundamental knowledge of botany and his great curiosity about the unknown. Thompson instructed him that he was to begin punctually at 2 pm each week from Friday 17th September, for which he would be paid 11/- [eleven shillings = $1–10] per session, a total of £6-12-00. By 13th December the family was living in *Lawson Terrace*, Powlett St, East Melbourne, and the lights were on!

4.2 Ormond College

At the time of McAlpine's arrival, The University of Melbourne had been established for just 30 years and although University land had been set aside for colleges in 1856, it was not until 1872 that the first College was built (Blainey 1957). Trinity was the first to take in students, while the second to open was Ormond College in 1881. Most others were established during the remainder of the decade.

The founders of the Melbourne colleges looked to the English model where colleges had traditionally provided most lecture courses and tutorials, although by the 1880s the traditional role of the Colleges had been taken over by the Universities, and the Colleges provided accommodation and tutorial backup. The latter model was proposed for Melbourne but the quality of teaching in the natural sciences at the University of Melbourne was so poor that many students avoided University courses, opting instead for Colleges tuition. This resulted in friction between the University and its Colleges (Blainey 1957).

In 1855, Professor Frederick McCoy[A1] was appointed to the Chair of Natural History. In Ireland, he had established his reputation as a geologist and palaeontologist, but as professor of natural history, he covered everything from anatomy and botany to chemistry, geology and zoology. In the first three decades, the University's resources were thinly spread especially in the sciences, and by 1883, Alexander Sutherland[A1], Headmaster of Carlton College, a member of the University Senate, and later University Registrar, was highly critical of McCoy (Blainey 1957). He argued that the importance of industry in Australia justified the teaching of science and claimed that from a scientific point of view, chemistry was not taught at all at the university and that the teaching of science generally was not worthy of the name. In consequence, in 1883, the University approved the conferring of science degrees, but did not frame the regulations for the 3-year degree course.

Meanwhile, in 1884, students at the University complained that McCoy taught botany in a lecture room and not in a garden (Blainey 1957). This set the scene for change and provided an opportunity for McAlpine. On 6th December 1884, McAlpine wrote to the Premier of Victoria, Hon. James Service, asking for a recommendation for a teaching post in biology at the University of Melbourne. The Premier (Service 1884) responded that while he had not met McAlpine, he had been mentioned to him by His Excellency Sir George Strahan, Governor of Tasmania, and wrote; – "I feel therefore that your testimonials, copies of which were forwarded to me, speak for themselves with regard to your qualifications for the position you seek."

In consequence on 18th February 1885, Dr. John Henry McFarland[A1], Master of Ormond College, appointed McAlpine to provide twice weekly practical classes in Biology coordinated to illustrate lectures given by Professor McCoy. The cost of these classes was advertised at one guinea [£1-1-0] per term plus a 5/- fee for the use of a microscope. There were also evening classes offered to Ladies and non-university members of the public for two guineas per term, microscopes included. By the end of 1886, McAlpine appears to have been giving lectures as well as practical sessions and was being paid per lecture. He was paid £3-10-00 [for 6 lectures, i.e. $11^{s}/8^{d} = \$1.18$ per lecture] in September and £4 in October [for 8 lectures, i.e. 10/-, = $1 per lecture]. The quality and value of his lectures and practical classes was greatly appreciated, and noted by the University hierarchy. McAlpine had presented McCoy with copies of the Botanical Atlases.

Also in 1886 the University framed the regulation for a 3-year science degree and appointed Professor W. Baldwin Spencer and Professor Orme Masson[A1] to the Chairs of Biology and Chemistry respectively in 1887, and they revolutionised the teaching of science. It emerged some years later that it was the quality of McAlpine's courses that encouraged the University to establishing the Chair to which Baldwin Spencer was appointed (Blainey 1957). In 1824 Sir John McFarland (1924) wrote to McAlpine acknowledging the importance of his work at Ormond and told him it was the quality of his courses that decided the University to establish the Chair of Biology.

4.3 Auld Po-Tarsh: Victorian College of Pharmacy

At the time McAlpine was seeking an appointment within the University, he sought support from Edinburgh for an application to lecture at the Victorian College of Pharmacy in Melbourne. Early in January 1885 he received a letter from R. Urquart (1885), Secretary of the Pharmacy Section of the School of Medicine, Edinburgh, stating that: – "Mr. Daniel McAlpine lectured in Botany and Natural History in this Institution for some time previous to his leaving the country. He is sure to give every satisfaction." This opened the way for what was possibly the happiest period of McAlpine's teaching career, especially in terms of the appreciation, respect and satisfaction it brought him. He lectured there from 1886 to 1911 (Sisson 1968). It earned him the respect and affection of his students and colleagues to which he responded with warmth and friendship (Anon 1932a, b, 1934).

Some insight into McAlpine's success as a respected academic and teacher is evident from the expressions of appreciation from past students who attended his lectures in biology at the Pharmacy College between 1885 and 1911 (Anon 1932a, 1934). Dr. B.L Stanton described him as "a youth with a white beard", who in rare degree compelled and retained warm affection among his students, colleagues and associates. He commented that an often heard little rhyme quoted by McAlpine was:

So long as life her term extends
The joy of nature never ends.

Stanton referred to the excellence of his two-volume *Botanical Atlas* as a teaching aid, his skill in the laboratory and his ability as a microscopist. He said that although McAlpine's brisk delivery and Scottish brogue could be a problem for some students, it was clearly also part of his charm. J.H. Gould described how he would write on the blackboard, then almost jump around to face the class to explain the importance of what he'd written, adding that the lectures were a joy. Several students referred to his pronunciation of "potash", which came across as "Po-tarsh". So much so that his nickname within the College became "Auld Po-Tarsh". Dr. C.J. Tonkin remembered his particular pronunciations of "proto-plarsm" and "Gorr-se", and how, when asked a question, he always began his reply with, "Weell, don't ye see—". He also said McAlpine was adept at controlling would-be disruptive students with wit and firmness, much to the amusement of the rest of the class.

A.R. Bailey described Old Po-Tarsh as upright, always in a frockcoat and top hat (Fig. 4.1), and always rushing with his coat tails flying. He said McAlpine made the story of the life cycle of ergot more wonderful and interesting than a three-volume novel, and took two lectures to cover it! He was an enthusiastic lecturer. His lectures

Fig. 4.1 *Auld Po-Tarsh*, explaining some intricacy of botany to students on a field excursion (Courtesy of the State Library of Victoria, Melbourne, La Trobe Australian Manuscripts Collection, MS 12772)

were clear and interesting, while the light he lit for botany was never extinguished. E.T. Church said he was privileged to attend McAlpine's classes, and held him in great respect and deep affection. He admired the keen interest McAlpine took in the progress of pharmacy generally. C.C. Wallis wrote that he loved McAlpine's charm of manner, Scottish wit and dignity, all associated with a sensitive and poetic nature, which showed through his lectures. He paid special tribute to McAlpine's friendship and the quality of his lectures and excursions.

In the field and on excursions he was untiring, and lost all track of time. J.H. Gould related how McAlpine was untiring on excursions to Blackrock and Cheltenham. With a string of students strung out behind, perspiring as they tried to keep up, McAlpine hopped about from flower to flower and tree to tree like a bird. It is clear that from Edinburgh to Melbourne his students admired, appreciated, valued and cherished him.

Following his appointment to investigate the cause and control of bitter pit, the staff, graduates and students of the Pharmacy College organised a farewell attended by 150 ladies and gentlemen at which they presented McAlpine with a gold watch in recognition of his 27 years of dedicated service to the College. They also presented Mrs McAlpine with a valuable vase. They both treasured their gifts and the affection and friendship they signified.

4.4 Research and Other Occupations 1884–1890

While at Ormond, McAlpine developed methods for the dry preservation of animal specimens for class purposes (McAlpine 1886a) and similarly the dry preservation of flowers and fungi in a manner, which retained their original shapes and colours

(McAlpine 1886b). He also published observations on the movement of detached parts of bivalve molluscs (McAlpine 1887a), and expanded this study in some detail (McAlpine 1887b, 1888). He collaborated with J.R. Remfrey (McAlpine and Remfrey 1889) in investigating differences in the anatomy of petioles as a possible means of identifying species of eucalypts. His next interest was in aspects of the biology of copper head snakes. He found two copperheads at Oakleigh near Melbourne, one of which he killed by decapitation, the other he captured alive. On dissecting the decapitated snake some hours after it was killed, he was surprised to find its heart still beating, which it continued to do so for about 8 h after decapitation. He published three papers on his study of the snakes. One on a parasitic fluke (McAlpine 1890a), a second on a nematode in the snake's stomach (McAlpine 1890b), and the third on movements of the heart, in and out of the snakes' bodies (McAlpine 1890c).

His interest in Pharmacy was also maintained. Eleven months after arriving in Australia he published an article in the *Australian Journal of Pharmacy* under the title, *Modern Botany in its Relation to Pharmacy* (McAlpine 1885) in which he drew attention to the need to advance beyond the old descriptive botany and learn about the physiology of plants and the chemical nature of their properties. He also took an interest in the cultivation of medical herbs (McAlpine 1886c). These articles followed his appointment as Lecturer in Botany at the Victorian College of Pharmacy. As time passed, he also lectured at the School of Horticulture, Richmond established in 1891. He gave the inaugural monthly public lecture, *Botany in its Relation to Horticulture,* in 1892 (Anon 1893). These lectures were additional to those delivered as part of the curriculum, were open to the public, and given by selected experts. Baron Ferdinand von Müller delivered the second lecture on, *Our Indigenous Plants in relation to Horticulture*, with others given by Charles French Snr. on entomology and W.R. Guilfoyle on *British Botanical Gardens and their Conservatories* (Anon 1893). McAlpine also acted as examiner in various institutions such as the Melbourne Veterinary College (White 1986).

Up to the late 1880s, there was no published evidence of McAlpine having a particular interest in fungi or plant diseases. His interest in natural history was almost unlimited. As a lecturer he was expected to cover a broad range of sciences, a common expectation of lecturers at the time. During the late 1880s, however, calamitous plant disease epidemics made him aware that disease epidemics in agricultural crops such as cereals, in addition to the potato, also caused grave consequences. As a keen naturalist with a love of fieldwork, he began to take an interest in the rust, bunt and smut fungi, which were devastating crops and which were common on native grasses and other plant groups.

4.5 A New Vocation Beckons

Within 2 or 3 years of his arrival in Australia the recognition of McAlpine's abilities as an authoritative natural scientist had spread beyond the lecture room. Alfred Deacon, then Minister for Agriculture and water supply in the Colony of Victoria and destined to be Prime Minister of Australia[A1] was one, Baron Sir Ferdinand von Müller[A1], Victorian Government Botanist was another, as was Dr. J.H. McFarlane, the Master of Ormond College who appointed McAlpine and was later Chancellor of the University of Melbourne. It was not surprising therefore, that when the various Colonial Governments were forced to take action against the ravages of cereal rust, McAlpine was identified as a man with valuable knowledge and skills that should be used. The following year he was appointed Vegetable Pathologist. In order to appreciate Daniel McAlpine's achievements, his reputation among his peers and his place in the history of Plant Pathology, it helps to remind ourselves of the history of the development of Plant Pathology worldwide during the nineteenth century. It defines the intellectual base from which he grew and helps understand the context of his development. In a sense, his education began in the age of Darwin, de Bary, Berkley and Kuhn, developed with Millardet, Ward, Sorauer, Eiksson and Arthur, and ended in the era of E.J. Butler, W. Brown, E. Gaümann and E.F. Smith.

4.6 Natural Sciences in Nineteenth Century Europe

Plant diseases may well have been at least an inconvenience from the time Eve took the first bite from an apple! As agrarian societies developed, and the growing of crops became increasingly important to sustain them, plant disease epidemics became more likely and the potential threats of destructive famines, capable of crippling affected communities emerged. In his epic book, *Advance of the Fungi,* E.C. Large (1940) described the impacts on society of calamitous plant disease outbreaks and how Plant Pathologist found ways to manage them. Carefoot and Sprott (1969) echoed Large's story in their book, *Famine on the Wind – Plant Diseases and Human History*. These books dealt mainly with individual plant disease epidemics, their contributions to human misery and history, as well as efforts to find effective remedies. In his *Introduction to the History of Plant Pathology*, G. C. Ainsworth (1981) emphasised the establishment and development of plant pathology as an applied science, which emerged in Germany in the mid nineteenth century. In doing so he acknowledged that there was valuable research and discoveries made during previous centuries, such as the invention of the microscope in the seventeenth century, and the use of lime, sulphur and copper salts to control some diseases, which emerged in the eighteenth century.

The outbreaks of the devastating disease of potatoes in the 1840s, now know widely as *Irish Blight*, which spread throughout England, Ireland, Scotland and Europe, focused public attention on the potential of a fungus to cause havoc in societies through wholesale destruction of a stable food crop. The impact of Irish blight was measured in the deaths of over a million Irish and thousands of others, added to which, the migration of 1.5 million Irish reducing the pre-famine population of 7.5 million to 5 million, post famine. Thousands of poor Scots also perished because of the potato blight famine, and with their Irish cousins set sail for the new world, especially America, Australia, Canada and New Zealand. Before the famine, these populations were already under severe economic stress and heavily dependant on the potato. In the Highlands of Scotland, the famine compounded the poverty and duress caused by the clearances. In consequence the famine focused public attention and awareness on the potential consequences to human welfare of disease in staple food crops. It is almost certain that McAlpine would have learned of this during the 1870s if not earlier. His illustration of *Phytphthora infestans* (Fig. 4.2), included in his *Botanical Atlas*, attests to this.

The English cleric, the Rev. M.J. Berkeley studied fungi as a hobby and was the authority in this field. His Belgian friend, Dr. Jean P.F.C Montagne, a physician who after retiring from Napoleon's Grand Army, devoted himself to the more peaceful study of cryptogamic botany. In 1845 Montagne sent Berkley drawings of what he had found during examination of blighted potato leaves. The drawings were of the sporangiophores of a fungus emerging from stomates around the perimeters of lesions in infected potato leaves. They published the drawings and a Latin description of the pathogen in the first issue of the *Journal of the Horticultural Society* in January 1846 (Berkeley 1846), with their conclusion that the disease was caused by a microscopic fungus. This was an unpopular view at the time, with more powerful opinion favouring a form of plant gangrene brought about in plants of poor health caused by a miasma. Berkeley cut thin sections of leaves where the sporophores were evident, and produced drawings of the fungal threads within the spongy mesophyll, showing sporophores leading out through the stomates. Berkeley and Montagne also recognised that the sporangia of the pathogen contained smaller particles, which they considered likely to be the infectious agents (Berkeley 1846). McAlpine drew heavily on such literature (Figs. 4.1 and 4.2).

Ainsworth (1981) claimed that from 1853, following publication of Anton de Bary's book, *Observation of the Smut Fungi and Plant Diseases caused by their Action* (de Bary 1853), there was general acceptance of the proposition that fungi were causal agents of plant diseases. While this view was held in some circles, it was not universal. The belief that moulds and other agents associated with decay and putrefaction arose through spontaneous generation was still widespread. From the 1840s, Louis Pasteur had questioned the theory of spontaneous generation and by the 1850s was convinced that putrefaction and fermentation occurred through similar processes. He therefore set out to determine what caused these processes, and throughout that decade ran a series of experiments, which proved that both fermentation and putrefaction were cause by living agents carried in the atmosphere (Large 1940). In spite of these findings, following his appointment as Vegetable Pathologist in 1890, McAlpine (1910) found that most farmers and politicians still believed poor weather caused disease and that the organisms found in dead tissue arose from it spontaneously.

While de Bary (1853) demonstrated the infectious nature of pathogenic fungi, described their capacity to grow in and damage plant tissue, and identified their propagules, which spread to and infected neighbouring plants, it was the publication of Julius Kühn's book, *Plant Diseases in Cultivated Plants, their Cause and Prevention* (1858), that marked the launch of Plant Pathology into the modern era (Ainsworth 1981; Large 1940). With an extensive background in agriculture and a

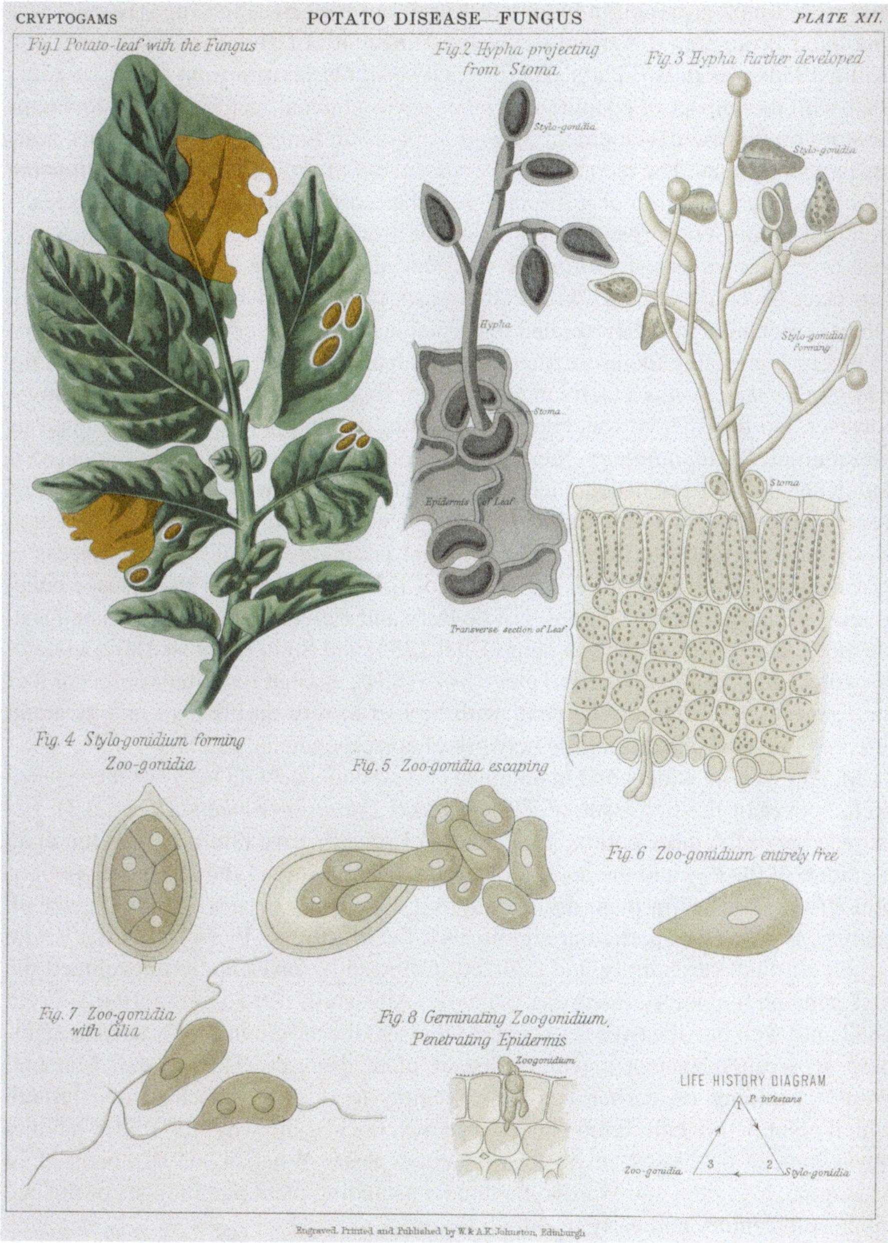

Fig. 4.2 McAlpine's life history of *Phytophthora infestans.* Based on de Bary. Botanical Atlas Vol. 2, Plate XII (1883) (Courtesy of the Royal Botanic Gardens Library, Melbourne)

PhD from Leipzig, based on his interest in diseases of beet and rape, Kühn[A1] was appointed Professor of Agriculture at the University of Halle in 1862. His book (Kühn 1858) was in two parts. The first, *General Observations on Plant Diseases* dealt with the impact of edaphic and other environmental factors such as low temperature on disease development as well as parasitic fungi, insects and other pests and parasitic plants. The second part *Specific Investigations of Diseases of Cultivated Crops* was a compilation of information on the most damaging diseases of cereals, including smuts, bunts rusts and ergot, and of beet, rape and other crops, much of it based on his own investigations. It was reprinted in 1859. Biographies of great European Mycologists and Plant Pathologists of the latter half of the nineteenth century often note that they studied under Kühn or de Bary or both.

It is not surprising therefore, that the first plant pathology journal, *Zeitschrift für Pflanzenkrankeiten* was a German publication. By the start of the twentieth century, however, North America was taking the lead in the field and in 1911 launched its own journal, *Phytopathology*. McAlpine was only a small boy when de Bary (1853) and Kühn (1858) published their books, but when he took up the position of Vegetable Pathologist in Australia in 1890, they were still important texts. Compared to a young graduate starting a career in plant pathology today, who has access to vast banks of knowledge in every branch of science bearing on the disease being investigated, McAlpine had copies of de Bary and Kuhn as general guides and article on specific diseases such as bunt (Tillet 1755) and Smut (Prévost 1807) of cereals, Irish Blight of potatoes (Berkeley 1847–1854), etc, but for a high proportion of the fungi and the diseases he dealt with he had to rely on his own initiative and research and advice from a wide network of correspondents.

McAlpine read widely and maintained correspondence with many of those listed on the cover of the first issue of *Zeitschrift für Pflantzenkrankeiten* (Fig. 6.3). His knowledge of the more common languages of science gave him access to the main literature of the day and the leaders of his discipline. Among the expanding pool of biologists interested in plant diseases during the closing decades of the nineteenth century, two were of particular significance for McAlpine. Pierre Millardet[A1] who largely through serendipity and curiosity followed by investigation developed the first general fungicide, *Bordeaux Mixture* (Ainsworth 1981; Large 1940). From 1882 until well into the twentieth century, this was the most commonly used general-purpose weapon against a wide range of plant diseases. The second, Marshall Ward[A1], building on earlier workers' attempts to discover mechanisms through which plant pathogenic fungi caused diseases, opened the way into understanding physiological processes involved in pathogenesis. Whetzel (1918) considered McAlpine, Millardet and Ward to be three outstanding plant pathologists of the late nineteenth century and early twentieth century.

As interest in teaching natural history evolved, it followed that organisms associated with diseased plants were studied and discussed, and in time plant pathology emerged from the miscellany of natural history towards the end of the nineteenth century. In Vienna in September 1890, an *International Phytopathological Committee* was established at a meeting of the *International Congress of Agriculture and Forestry* (Ainsworth 1981). The committee consisted of 44 members from 19 countries and included Daniel McAlpine. Paul Sorauer was the initiator and under his leadership the committee launched *Zeitschrift für Pflanzenkrankheiten*, the first journal dedicated solely to plant pathology. The following year Dutch members of the committee founded the *Netherlands Plant Pathology Society* and 4 years later published the first issue of their journal. In 1908 the *American Phytopathological Society* was established, and launched its journal, *Phytopathology* 3 years later.

4.7 Australia 1788–1890

European settlement of Australia began on the east coast in early 1788. One hundred years later, it was a devastating epidemic of wheat rust that led Daniel McAlpine into mycology and plant pathology. Within 2 years of attempting to grow wheat in the new colony of New South Wales [NSW] black stem rust was adding to the problem of growing crops of English wheat varieties which were generally unsuited to the Australian climate and cropping programs developed for the other side of the world. Such wheats were best suited to the milder coastal areas of the new colony, but there the mild moist regions of the coast also favoured rust. Cereal rusts may have been introduced accidentally or could have been already present on native grasses. During the first century of settlement there were ten serious epidemics of stem rust. The epidemic of 1889 caused exceptionally heavy losses and set in train a chain of events that drew McAlpine into an investigation, which had a worldwide impact (Large 1940).

During the previous century, however, considerable interest had been focused on the collection of fungi in Australia and in some cases on diseases caused by them. European explorers had visited Australia from at least as early as the seventeenth century or earlier and returned with collections of botanical specimens. Prior to settlement, collectors such as Sir Joseph Banks, botanist on the *Endeavour* with James Cook in 1770, as well as collectors on Dutch and French ships, sometimes returned to Europe with fungal epiphytes or parasites on plants they collected. These found their way to British and European mycologists who described, named and stored them in herbaria. Following settlement this pattern continued for many years, although most collectors were more selective in directing mycological specimens to specific experts of their day.

Early visitors referred to Australia by various names. The English called it Botany Bay and later New South Wales, while Europeans often referred to Australia as New Holland, some even as late as 1918. Between 1804 and 1859, however, Tasmania [van Demon's Land], Victoria, Western Australia [Swan River], South Australia and Queensland were established as independent colonies with their own Colonial Governments overseen by the Colonial Office in England. In 1901, these colonies were amalgamated as States, within the Commonwealth of Australia, each with its own State Government. Locations of many of the early collections still bear the old names.

Robert Brown, who accompanied Matthew Flinders during his 1801–1805 circumnavigation of Australia, made the earliest know collections after 1788. Twelve species were described and named. From then until 1890 most collections were sent to someone in Europe for identification, usually according to the nationality of the collector. Most sent by British settlers went either to M.J. Berkley, M.C. Cooke or G. Massee and were housed in the Kew Herbarium (Cooke 1892). Such collections included those by Cunningham (Tasmania 1816–1831), Gunn, (Tasmania 1832–1850), Drummond (Tasmania 1839), Preiss (Western Australia 1839–1840) and von Müller (Victoria 1847- c.1890). Von Müller collected extensively in many parts of Australia including the Northern Territory and sent specimens to European mycologist.

Collectors were much more active in some colonies than others, there being no records of specimens sent overseas from NSW. An investigation by John Walker (2014) found that from 1890 until 1905, Nathaniel Cobb collected many specimens and published illustrated articles in the *Agricultural Gazette of New South Wales* and in his *Letters on the Diseases of Plants.* He identified most of his own specimens. Walker found no evidence of Cobb's collaboration with McAlpine or anyone else, while other NSW collectors such as Musson and Carne did send specimens to McAlpine. Cobb's interests in taxonomy related mostly to nematodes and very few mycological specimens collect by him were lodged in the NSW Department Herbarium [DAR].

Queensland was separated from NSW as an independent colony on 6th June 1859. Three men made substantial contributions to mycology and plant pathology in Queensland over the following 40 years. Dr. Joseph Bancroft[A1] is considered the first to investigate a plant disease in Queensland, and probably the first in Australia, when in 1874, he diagnosed a destructive vascular wilt disease of sugar bananas, now known as Panama Disease (Bancroft 1876). Simmonds (1966) claimed that Bancroft's was the world's first report of this disease. Bancroft concluded that the best means of controlling Panama Disease was the selection of pathogen-free planting stock, evident by the lack of discolouration in the vascular system of affected mother stock (Bancroft 1876). His recommendation is still followed, although the methods of securing such planting stock using tissue culture techniques has greatly improved its reliability. Bancroft made a number of important advances in both animal and plant health.

Frederick Bailey[A1] settled in Brisbane in 1861 where he established a seed store and collected specimens of indigenous Australian plants for sale to British and foreign botanical nurseries. He was later appointed botanist to a Government Board (Bancroft 1876) established to deal with serious animal and plant health problems. The same Board recorded Bancroft's work on Panama disease of bananas. While Bailey is best known for his publications of phanerogams, he was also interested in cryptogams and send numerous specimens for identification and comment to M.J. Berkley, M.C. Cooke, and G. Massey and after 1890, to McAlpine. Most pre 1890 specimens were incorporated into Cooke's *Handbook of Australian Fungi* (Cooke 1892). Cooke later lent the Kew Herbarium's collection of Australian fungi to McAlpine, to assist him prepare his *Systematic Arrangement of Australian Fungi* (McAlpine 1895).

Henry Tryon[A1] arrived in Brisbane in 1882, where he became a keen plant collector which led into a fascination with the insects associated with them. These he collected, studied and described, building up a substantial collection. This resulted in his appointment to the Queensland Museum in 1883 where he engaged in entomological research. The following year he was appointed Government Entomologist and in 1891, his designation was changed to Government Entomologist and Vegetable Pathologist. In his latter role he studied the interactions of parasites with their hosts and devised control measures based on what he had learned. He was the first to report Irish Blight of Potatoes in Australia. He played an important role in the introduction of *Cactoblastis cactorum* as a biocontrol agent to combat the unbridled invasion of thousands of acres of valuable land by the Prickly Pear, *Opuntia vulgaris*. This was the first successful biological control of a pest in Australia. In 1922, the Queensland Government enabled Tryon to appoint four new members of staff, including the first appointments of B.Sc. graduates, one of whom was J.H. Simmonds (1991). It is evident that at least one of the reasons McAlpine was seldom invited to help solve problems in Queensland was that the above three largely self-educated plant pathologists managed on their own.

In Victoria, prior to McAlpine's appointment as Vegetable Pathologist, collections of mycological specimens, including those of von Müller, were sent overseas for identification. From 1849, when von Müller was naturalised, his ambition was to write a flora of Australia. Realising the enormity of the task he recruited a large network of collectors including 225 women (Maroske and Vaughn 2014), one of whom was Miss Mary Campbell who collected widely mostly in Victoria and southern New South Wales. In early 1888 she was employed by the Victorian Department of Agriculture to investigate the 'hop-spider' [red spider mite] which was damaging hop production in the hop fields around Bruthen. She found that the mite was favoured by the over watering and poor drainage of hopfields (Argus 1888). Von Müller sent most collections of fungi overseas to Cooke and others for identification and eventually most found their was back to McAlpine during his preparation of his systematic arrangement of Australian fungi (McAlpine 1895). It appears that Miss Campbell was offended by McAlpine's failure to acknowledge her collections in his monograph (McAlpine 1895). Her reaction may have been heightened by disappointment at not being appointed vegetable pathologist (Dr. Tom May, pers.com.).

While McAlpine lectured to Pharmacy students about fungi that caused health problems in humans and to Veterinary students regarding animal health problems, it was Professor A.J. Ewart that first taught plant pathology, not by that name but integrated within courses in general botany for agricultural science and science students. It is evident that this was also the case in Queensland (Simmonds 1991), and it was probably the same in most if not all Australian Universities until such time as lecturers were appointed specifically to teach plant pathology.

4.8 Formal Courses in Plant Pathology

According to Ainsworth (1981), in Europe during the eighteenth and nineteenth Centuries most knowledge of mycology and plant diseases was gained by self-taught amateur naturalists, teachers of natural history or students of medicine. While the subjects taken and passed by McAlpine are known, the content of courses he took is not. For much of the nineteenth century subjects such as agriculture, biology, botany, zoology and geology were all components of Natural History. Perusal of the *Biological Atlas* compiled by the McAlpine brothers in 1880 suggests that at the time mycology was taught as a component of cryptogamic botany. However, as knowledge increased natural history split into more defined components of animal vegetable and mineral.

We may not know what McAlpine was taught as an undergraduate but we do know his lecturers, especially Huxley and Thiselton-Dyer inspired him to excel. This led to his becoming a lecturer, and to the realisation that students needed readily available texts in English. The terms *mycology* and *mycologist* may have been familiar to him, as they had been used in England from the mid 1830s (Ainsworth 1876), however, the term for the study of plant diseases appears to have emerged in Europe 60 years or so later when Dutch and German biologists began using their equivalent terms for phytopathology and phytopathologist. Dr. Jack Simmonds (1991) stated that at the time of his own appointment to the Queensland Department of Primary Industries in 1921, he had not heard the terms "plant pathology" or "plant pathologist" and no such subject was taught at the University of Queensland. It was usual for fungi and plant diseases to be included in courses on general botany. No course named "Plant Pathology" appears to have been taught in Australia much before the mid 1920s.

Dr. Linden Gillbank (2010) investigated the teaching of plant pathology at Melbourne University. It began following Professor A.J. Ewart's appointment, to service agriculture and to educate agricultural scientists. As Government Botanist Ewart advised on problems associated with economically important plants. As Professor of botany he included lectures on pathogens and plant diseases in courses on general botany and agricultural botany. This was not unusual at the time. 'Fungi, with special reference to injurious and useful forms' and 'Methods of combating diseases in plants' were included in the subject description for Agricultural Botany given in the *Melbourne University Calendar* for 1907. The *Calendar* for 1912 included the classification and morphology of Phycomycetes, Ascomycetes, Basidiomycetes and Lichens; and for agricultural botany, 'The outlines of Plant Pathology, with special reference to the commoner diseases of cultivated plants due to bacterial and fungoid parasites and to those diseases of animals which are due to Fungi'. Ewart also taught the first trickle of Botany Part II students about pathogens and plant diseases and organized a separate course on plant pathology for 5th year veterinary students, emphasising diseases of pasture plants. He supervised postgraduates, including Ethel McLennan who was a meticulous researcher, research supervisor, lecturer and botanical and mycological artist. For three decades from 1926 she shaped the development of plant pathology and mycology in the Botany School.

McLennan taught Mycology and Plant Pathology as independent subjects until 1956 (Gillbank 2008, 2010). She lectured to generations of science and agriculture students and trained an enduring dynasty of plant pathologists such as D.B. Adam[A1], N.J. Flentje[A1], G.M. Weste[A1] and many others (Parbery 1989). Professor Ewart may have sparked Ethel McLennan's lifelong fascination with fungi, but he denied her the opportunity and benefit of meeting McAlpine. McLennan's thesis 'Endophytic Fungus of *Lolium*' earned her the University's 47th D.Sc. degree, which was conferred in April 1921. It was the University's first D.Sc. degree awarded for mycological research, and possibly the first from an Australian University. Professor Ewart considered her work on the *Lolium* endophyte the most important pre-World War 1 mycological work in Australia (Ewart 1921), while ignoring the enormous contribution made by McAlpine.

References

Ainsworth GC (1976) Introduction to the history of mycology. Cambridge University Press, Cambridge, UK, 359 pp

Ainsworth GC (1981) Introduction to the history of plant pathology. Cambridge University Press, Cambridge, UK, 315 pp

Anon (1893) Monthly lectures delivered at school of Horticulture, Burney, by various specialists during 1892–1893. Government Printer Melbourne No. 826, for Department of Agriculture, Victoria, on behalf of School of Horticulture Est. May 1891

Anon (1932a) Death of Daniel McAlpine. Aus J Pharm 13:933, 1003–1005, 1037, 1096

Anon (1932b) Daniel McAlpine – Obituary. Chemists and Druggists of Australasia 47:167

Anon (1934) Obituary. Australas J Pharm 15:473

Argus Newspaper (1888) Article about the vegetable production commissioners' report on their visit to Gippsland [22nd Feb 1888]

Berkeley MJ (1846) Observations botanical and physiological on the potato murrain. J Hortic Soc 1:9–34

Blainey G (1957) A centenary history of the University of Melbourne. Melbourne University Press, Melbourne, 220 pp

Carefoot GL, Sprott ER (1969) Famine on the wind – plant diseases and human history. Angus & Robinson Publishers, London, 222 pp

Cooke MC (1892) Handbook of Australian fungi. Williams & Norgate, London, for the Departments of Agriculture in Melbourne, Brisbane, Sydney Adelaide, Hobarton

De Bary A (1853) Untersuchungen über Die Brandpilze und die durch sie Verursachten Krankheiten der Pflanzen. Müller, Berlin, 144 pp, 8 plates

Donaldson J (1884) Letter [23-5-'84] written to the Governor of Tasmania, with a testimonial to McAlpine's ability as a natural science lecturer

Ewart AJ (1921) Speech published in the Proceedings of the Australian association for the advancement of science conference, Melbourne, Jan 1921

Gillbank LR (2008) Fungi – Mycology and plant pathology. Unpublished notes on the History of the School of Botany, University of Melbourne, 12 pp

Gillbank LR (2010) From system garden to scientific research: the University of Melbourne's School of Botany under its first two professors (1906–1973). School of Botany, University of Melbourne, 40 pp

Gippsland Times (1884) Report of the arrival of the SS Gulf of Carpentaria in Hobart on 23rd July 1884

Gippsland Times (1885) Report of the Loss of the SS Gulf of Carpentaria. Wednesday 16th Sept 1885, p 3

Hooker SJ (1884) Letter [26-5-'84] written to Mr. Abbott, curator of the Hobart Botanical Garden, with a testimonial to McAlpine's abilities

Kühn J (1858) Der Krankheiten der Kulturgewächse, ihre Ursachen und ihre Verhütung, Berlin

Large EC (1940) The advance of the fungi. Jonathan Cape, London, 488 pp

Maroske S, Vaughn A (2014) Ferdinand Müller's female plant collectors: a biographical register. Muelleria 32:92–172

McAlpine D (1885) Modern botany in its relation to pharmacy. Aus J Pharm 31:7–9

McAlpine D (1886a) Dry preparation of animal specimens for museums and teaching purposes. Vic Nat 3:14–22
McAlpine D (1886b) The dry preservation of flowers, fungi, etc., in their natural shapes and colours. Vic Nat 3:117–126
McAlpine D (1886c) The cultivation of medicinal herbs. Chemists and Druggists, London
McAlpine D (1887a) Notes from the biological laboratory, Ormond College part 1- observations on the movement of detached Gills, Mantle-lobes, labial palps and foot in Bivale Mollusks. Trans R Soc Vic 15:139–150
McAlpine D (1887b) Notes from the biological laboratory, Ormond College part 2- observations on the movement of Detached Gills, Mantle-lobes, labial palps and foot in Bivale Mollusks. Trans R Soc Vic 15:174–204
McAlpine D (1888) The movements of the Edible Mussel and its various parts. Vic Nat 45:9–11
McAlpine D (1890a) Remarks on a fluke, parasitic in the copper head snake (*Hoplocephalus superbus* Günth.). Proc R Soc Vic NS 3:40–43
McAlpine D (1890b) On a nematode found in the stomach of a copper head snake. Proc R Soc Vic NS 3:36–39
McAlpine D (1890c) Observations on the movements of the heart of the copper head snake, in and out of the body. Proc R Soc Vic NS 3:27–35
McAlpine D (1895) Systematic arrangement of Australian fungi together with host index and list of works on the subject. Department of Agriculture/Government Printer, Victoria/Melbourne
McAlpine D (1903) Report of the vegetable pathologist for 1902–1903. J Dep Agric Victoria 11:1–7
McAlpine D (1910) Twenty years of plant pathology in Australia. Report of the vegetable pathologist for 1907–1910. Department of Agriculture, Victoria, pp 56–61
McAlpine D, Remfrey JR (1889) Transverse sections of petioles eucalypts as aids in the determination of species. Proc R Soc Vic NS 2:1–63, 7 Plates, 40 Figs
McFarlane J (1924) Letter to McAlpine [21-10-"24] regarding his influence on the establishment of the Chair of Biological Sciences at Melbourne University
Parbery DG (1989) Ethel Irene McLennan 1891–1983: pioneer teacher of mycology and plant pathology in Victoria. Australas Plant Pathol 18:47–56
Prévost B (1807) Mémoire sur la cause immédiate de la cause de carie ou charbon de blés, et plusieurs autres maladies des plantes et sur les preservatifs de la carie. Paris. English translation, G.W. Keitt, Phytopathological Classics 6, 191 pp, 1939
Service J (1884) Premier of Victoria's letter [13-12-1884] responding to DM's request of 6th December '84 for support in obtaining a teaching post at the University of Melbourne
Simmonds JH (1966) Host index of plant diseases in Queensland. Queensland Department of Primary Industries, Brisbane, 111 pp
Simmonds JH (1991) A short history of the early days of plant pathology in Queensland. Qld Nat 31:4–12
Sissons ATS (1968) History of the Victorian college of pharmacy. Aust J Pharm 49:497
Thompson JA (1884) Letter written [9-9-1884] to McAlpine setting out the terms for teaching at Kew high school
Tillet M (1755) Dissertation sur la cause qui corrumpt et noircit les grains de blé dans les épis: et sur la moyens de prevenir ces accidents. Bordeaux. English translation, H.B. Humphrey, Phytopathological Classics 5, 191 pp, 1955
Urquart R (1885) Testimonial [2-1-1885] for McAlpine from Urquart, secretary of the pharmacy section of the school of medicine, Marshall Street, Nicolson Square, Edinburgh
Walker J (2014) Correspondence sent to the author [5-2-'14]
Walters SM (1989) Introduction to the Bracken books republication of The Botanical Atlas, a Guide to the Practical Study of Plants (Part 1 and 2 Combined) by Daniel McAlpine, Best Seller Publications Ltd., Princess House, 50 Eastcastle Street, London. 136 pp, ISBN 1 85170 255 5
Wedge E (1984) Letters [4th to 7th May-1984] to D.G. Parbery, La Trobe Collection
Whetzel HH (1918) An outline of the history of phytopathology, with 22 portraits. W.B. Saunders Company, Philadelphia/London, 130 pp
White NH (1986) McAlpine, Daniel (1849–1932). In: Australian dictionary of biography, vol 10. Melbourne University Press, Melbourne, pp 193–194

Chapter 5
Vegetable Pathologist

5.1 Devastating Epidemics

Following the devastating epidemic of wheat rust in 1889, mounting public concern over losses resulted in loud calls for action. Black stem rust of wheat had been a problem in Australia from soon after settlement in 1788. In spite of information about the cause of wheat stem rust being around for over a century (Ainsworth 1981), there were farmers and politicians who blamed it on unfavourable weather and soil conditions. By then, de Bary's *Morphology and Biology of Fungi* (de Bary 1887) was available in English, and the better educated were aware that the rusts were wind-borne. But they were a minority. One farmer blamed fermentation of the roots, believing the red powder on stems and leaves was merely iron oxide shed from rusty windmills and other sources. Others believed that rust infections occurred in the same way as seedling-infecting bunts and smuts and that seed dressings would control them. In the meantime, between about 1840 and 1860, stem rust pushed wheat-growing inland from the relatively mild moist coastal belt and alluvial river valleys into the drier, hotter interior, quite unsuited to English wheats (Henzell 2007) (Fig. 5.1).

Eventually, during the closing decades of the nineteenth century, several severe epidemics forced authorities to face the need for action. The rust epidemic of 1889 was estimated at the time to have cost the wheat industry between £2,000,000 and £3,000,000 [between $300,000,000 and $500,000,000 in 2012 Australian dollars]. This alerted the Colonial Governments to the need for action at the political level to deal with it. Alfred Deacon[A1], Victorian Minister for Agriculture and water supply, had concluded there was a need to appoint a Vegetable Pathologist to investigate crop diseases and find means of controlling them. Deacon wrote to the Hon. Walter Madden[A1] advocating the appointment of Daniel McAlpine as Vegetable Pathologist. Madden, elected to the Legislative Council of Victoria as member for the Mallee, had seen the devastation caused wheat rust and by rabbits and no doubt also about the scourges of smuts and bunts. Deacon was aware that over the latter part of the nineteenth century most agriculturalists and politicians believed plant diseases were

D.G. Parbery, *Daniel McAlpine and The Bitter Pit*,
DOI 10.1007/978-3-319-09552-3_5

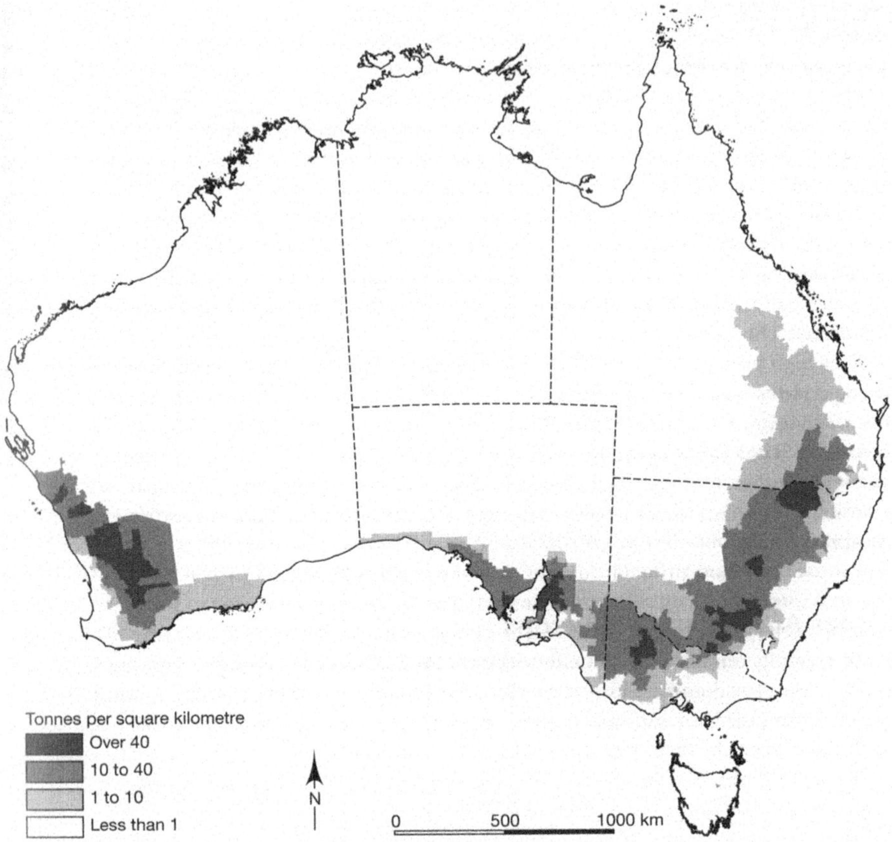

Fig. 5.1 Map of the wheat growing areas of Australia. The unshaded eastern coastal strip was abandoned during the late nineteenth century. Ted Henzell (2007) *Australian Agriculture – Its History and Challenges*, Fig. 1.8, p 33 (Permission of CSIRO Publishing, Melbourne)

caused by unfavourable weather, but he believed otherwise. The following year on 12th May, with strong support from von Müller, the Victorian Government appointed Daniel McAlpine Consulting Vegetable Pathologist on a salary of £200 *per annum*. It was the first appointment of a government plant pathologist anywhere in the British Empire. The only earlier recorded appointment was in the USA in 1885.

In the meantime, the Australian Association for the Advancement of Science (AAAS), established by Baron Ferdinand von Müller in 1888, appointed a Steering Committee to formulate a proposal for a preliminary program of experimental work aimed at reducing the rust menace. This was for consideration at the First Rust-in-Wheat Conference, planned for Melbourne in March 1890. It is evident that prior to this (Anon 1890; McAlpine 1889), McAlpine's interests and abilities had come to the attention of influential men such as the Hon. Alfred Deacon and Baron von Müller (Chairman of the Steering Committee). The other members of the

Steering Committee were Alfred Pearson, Professor W. Brown and Daniel McAlpine (Fish 1976). The program put forward by the Steering Committee was adopted at the Melbourne Conference. Members of the AAAS Committee responsible for organising the actual conference were J.H. Maiden FRS[A1], D. McAlpine, C.A. Topp[A1] and F. Wright, with A.N. Pearson[A1] as Secretary (Gillbank, pers. comm.). No information is available to me regarding F. Wright or W. Farmer unless the latter were P.W. Farmer MB, with whom McAlpine had had an earlier association (McAlpine and Farmer 1893).

Following the Melbourne meeting, Mr Smith Ellis, claimed he had developed a seed dressing that controlled rust and that he would reveal its composition if he received a reward of £10,000. His claim caught the attention of the Minister for Agriculture who, believing the treatment should be tested, agreed to pay the reward if it proved effective. Ellis had arrived from Yorkshire, England, in 1852. He came with a farming background and soon established a wheat farm on the Mornington Peninsula, between Port Phillip Bay and Westernport Bay. Within 6 years he ceased wheat farming and turned to sheep because stem rust had made wheat farming unprofitable. However, he had concluded that wheat rust infected seedlings from spores carried on the grain and he conducted trials for many years in an effort to find a satisfactory seed dressing to control it.

When the reward was offered, Mr Ellis said he would reveal the secret remedy once he secured the money. In view of the cost of rust epidemics, the Minister was persuaded that Ellis should be assisted to demonstrate the efficacy of his treatment. Ellis's claim and the offer of the reward were soon announced in the press, generating considerable interest, debate and dissent. An interview with Sir Henry Parkes[A1], Premier of New South Wales, was reported in the *Kilmore Free Press* (1890). Sir Henry thought the reward would be money well spent if the alleged control measure worked and he pledged his Government's support, providing the other Colonial Governments also contributed. The reporter, Mr Gillies, also interviewed the Queensland Premier who agreed the reward was a worthwhile idea, but suggested it should be made available to the first person to discover a successful remedy, as judged in an open competition. The *Gippsland Times* (1890) reported Ellis's claim and the Minister's support for testing its value. It also reported that the Victoria's Government Chemist, Mr A.N. Pearson, doubted such a remedy would be effective. In May, the *Colac Herald* (1890) also ran an article in which both Professor W. Brown and Alfred Pearson argued strongly against the likelihood that such a remedy could work.

The Minister, however, provided Ellis with the opportunity to demonstrate the effectiveness or otherwise of his treatment by arranging with William Guilfoyle[A1], Director of the Botanic Gardens in Melbourne, to make a plot of land available to Mr Ellis for the purpose. Ellis took up the offer and proceeded to plant his treated seed. But fate intervened. On the morning of 30th July, it was reported in the Argus Newspaper (1891) that staff at the Botanic Gardens had seen Ellis fall while weeding his plot and rushed in to render assistance, only to find him dead. No one ever collected the reward. This tale highlights two challenges facing the prospective plant pathologists: that of relieving agricultural industries of the financial burden of plant disease, and that of educating farmers about the causes of diseases, how they were borne and how best to control them.

McAlpine's appointment as the Victorian Vegetable Pathologist was made on 12th May 1890. Three months earlier, on 17th March, the *Brisbane Courier Mail* reported that in early March McAlpine had been invited to join the AAAS Steering Committee. His appointment was made 2 months after the First Intercolonial Rust-in-Wheat Conference opened on 11th March 1890, but McAlpine had been unable to attend, possibly because he had other commitments or had no official standing. Prior to the conference, however, under von Müller's chairmanship, McAlpine, Brown and Pearson had developed the set of recommendations for experimental work, which were adopted with few modifications. Under their program, officers of the various Colonial Departments of Agriculture were to conduct research programs determined at the conference. The experiments were usually run in conjunction with selected farmers. Following his appointment McAlpine took responsibility for the overall coordination and supervision of the program.

McAlpine attended the second conference held in Sydney in 1891 (Fig. 5.2), the third in Adelaide in 1892, the fourth in Brisbane in 1894, and he chaired the fifth held in Melbourne in 1896. He concluded that the only way forward was to follow Farrer's example and breed wheats suited to Australian conditions (Fish 1976). Farrer[A1]

Fig. 5.2 Delegates at the Intercolonial Rust Conference. Sydney, June 1891. Standing; D. McAlpine (Vic.), W. Farrer (NSW), H. Kelly (SA), G. Inglis (SA), N. Cobb (NSW), H.C.L. Anderson (NSW), Seated: Mc. Knight (Vic), F.W. Shelton (Qld), Hon. Sydney Smith (NSW), P. McLean (Qld), A.N. Pearson (Vic) (Courtesy University of Sydney, Cobbitty images collection)

attended the Rust-in-Wheat Conferences (Wrigley 1981) and it is probable that McAlpine and he met in Sydney in 1891. They co-operated closely over the following decade. The three key scientists who worked towards reducing wheat rust and placing the industry on a sound footing were William Farrer[A1] breeder, Daniel McAlpine plant pathologist and Frederick Guthrie[A1] chemist.

McAlpine strongly influenced the direction and planning of the work and was involved as a researcher. He established his own experimental plots, each of an acre of land on farms at Port Fairy, Myrniong, Nhill, Swan Hill and Waitchie (McAlpine 1900b), which were subject to a wide range of climatic conditions. Each site was within easy reach of a railway station. Similar programs were run in the other colonies. For example, William Angus claimed to have assisted with the Southern Australian trials (Daniels 1979), his claim being supported by McAlpine's having given him a signed copy of *Rust of Australia*, now in possession of Dr. Vyrna Beilharz.

5.2 Founding Australian Plant Pathology

McAlpine's appointment provided him with a title and part salary only. In spite of minimalist support he pioneered plant pathology in Australia, working under difficult circumstances. He purchased his own equipment, including a microscope, and established a laboratory in his own home at 22 Armadale Street, Armadale, where he worked for much of his professional life. In 1900 he was provided with an assistant, Mr G.H. Robinson, which helped enormously. Prior to this, Robinson had been a senior horticultural student and gold medalist from Longerenong Agricultural College, Victoria and had helped McAlpine (1900d) in an honorary capacity prior to his appointment, after which Robinson enabled him to establish his herbarium, which by the end of 1908 contained 12,000 specimens. In 1906 McAlpine had been given an office at 422 Flinders Street Melbourne (McAlpine 1910b).

The 1900 Vegetable Pathologist's Reports (McAlpine 1900c, d), which refer to some activities of 1899, demonstrates the extent of McAlpine's activities and the progress he made during the first decade of his appointment. The volume, breadth and variety of his work can be gauged from both his own list of his publications (McAlpine 1910a), and his annual reports. For example in the 34-page annual report for 1900, he cites 700 specimens sent to him for diagnosis and advice, all of which he dealt with. The diseases, which occupied most of his time, are listed and then described at some length, several illustrated by C.C. Brittlebank who was freelance at that time (Figs. 5.3 and 5.4).

Wheats bred for resistance to rust, improved milling qualities and resistance to shattering and lodging were tested at five locations: Port Fairy, a south coastal district, and four locations in the drier inland Mallee and northern Wimmera regions of the state. He noted that for the first time a Sunshine harvester was used to strip the trial plots, each of one acre, and that the resulting harvests were much more accurate than previously obtained because there was less waste. Some plots were up to three acres. He noted that of the 55 new varieties under trial, 16 were to be retained

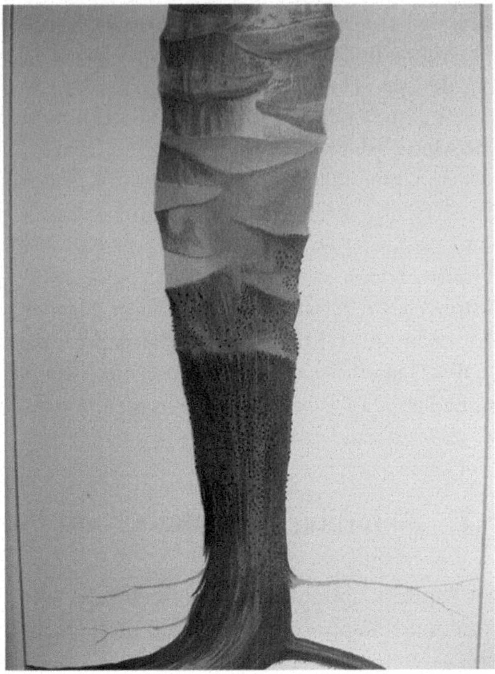

Fig. 5.3 Black Leg of Cauliflower by C.C. Brittlebank. Vegetable Pathologist's Report, Department of Agriculture Victoria for 1900 (Permission Department of Environment and Primary Industry, Victoria)

for further testing alongside new varieties sent by William Farrer. Trials aimed at finding varieties best suited to the Mallee were conducted at the Mallee Experiment Station. Here, 32 plots were established on 24 acres testing seeding rates and rust resistance. Mr A.N. Pearson conducted manurial experiments, which were reported separately. A smaller number of trials were conducted at Nhill, also in the Mallee. Nine varieties were tested at Swan Hill on the edge of the Mallee, and the fifth location was on the Brittlebank brother's farm at Myrniong on the Central Highlands. Here a flood wiped out the April sowing 3 weeks after emergence, however, the following sowing on 23rd June, harvested 23rd–25th December, was successful.

Flag smut of wheat had increased in importance since 1892 and various treatments were trialled to control it, including the use of clean seed, crop rotation, improved soil drainage and hot water treatment of seed, all of which provided a measure of control. In 1898, stinking bunt had cost the Victorian wheat growers £50,000 [2012 equivalent ca $18,000,000]. In 1899 and 1900, whereas the incidence of stinking bunt in the control plots were 33 and 76 % respectively, plots raised from seed dusted with or dipped in copper sulphate or certain other seed dressings were completely free of it. McAlpine referred to the increasing seriousness of bitter pit in apples and problems with other diseases for which he investigated control measures. These included black spot of apples, mould [downy mildew] of tobacco, club root and blackleg of crucifers, leaf curl, shot-hole and rust in stone fruits, and 'the grass destroying fungus'. The latter is rarely seen today, largely because of McAlpine's keen observation and advice.

During the closing decades of the nineteenth century, farmers had become concerned over a disease which was destroying large acreages of grassland in regions around the Ballarat district. McAlpine (1906b) identified the pathogenic fungus *Hypochnus fuciformis* which invaded the young foliage and killed it. Because the dead pasture took on a pinkish colour, the disease was commonly known as 'pink disease'. It infected a range of pooid grasses, but mainly pasture species such as *Festuca rubra* and *Lolium perenne* (Stalpers and Loerakker 1982). McAlpine noticed, however, that grass growing around cowpats and urine patches was healthy and lush. After conducting trials in which he treated affected areas with a range of nitrogenous fertilisers at various rates, he recommended dressing affected areas with 2 cwt per acre of ammonium sulphate (McAlpine 1900d). This was so effective that few if any of today's farmers or plant pathologists have ever heard of the disease or the pathogen. In spite of my interest in grass and pasture diseases spanning over 50 years, I have seen only one specimen, and that was in the mid 1960s.

Whereas in 1890 many Victorian farmers blamed the weather for disease outbreaks and were sceptical about the roles of infectious agent, by 1900 McAlpine had convinced them otherwise. He noted that in 1900, increasing numbers of farmers and orchardists were attending his lectures illustrated with lanternslides of diseases, the pathogens that caused them and sections of infected tissue with the pathogens in place. He also recorded the increasing numbers of producers who were visiting his various experimental sites to see the effects of various control measures. It is especially interesting to note that in the case of black scab of apples, McAlpine was testing the effectives of control treatments developed by orchardists themselves who then asked him to compare their effectiveness against his own treatments, which he did.

As busy as he was, he was always prepared to assist growers in their own endeavours to overcome disease problems. In his 38-page treatise on fungus diseases of cabbage and cauliflowers in Victoria and their control (McAlpine 1901), he explained that the growers themselves had come to the conclusion that they had to do what they could to get rid of infected trash in the fight against the ravages of blackleg. At their request he accompanied them on an inspection of a tar-fired rotary furnace designed for the disposal of garbage. He concluded that its use would be prohibitively expensive unless it could be set up as a co-operative dispenser of waste, but advised that the immediate removal of infected plants as well as other methods would mostly prove more practicable. At the end of the paper he added a two-page appendix, which included a report from J. Monash, best known as Sir John Monash, soldier and engineer, stating that the design of the furnace was sound and that similar rotary furnaces were proving useful for the disposal of garbage.

During 1904 McAlpine (1904c) also processed 1,179 specimens sent from every state in the Commonwealth. As well as responding to each of these enquiries, he published 15 papers. In the Victorian Fruit Inspector's Report for 1910, J.C. Turner (1910) provides insight into the reasons for this increase in the number of inquiries since 1900. Victorian apple exports began in 1891, the total value of fruit exported being £815. By 1909 this had jumped to £102,000, a 124-fold increase (over 12,415 %). Also, in the 2 years from 1905 to 1907, the area under fruit trees expanded from 47,312 acres to 61,927 acres, an increase of 31 %. McAlpine (1910e) also drew attention to the rapid expansion of agriculture, the increasing numbers of new crops and the new diseases that appeared. He continued his lifelong habit of reading and corresponding widely, which helped fulfill the need to devise and keep abreast of improved methods of disease control. In addition to fieldwork he continued to deal with twenty specimens for identification, diagnosis and advice every week McAlpine (1910d). During the years 1907–08, 1908–09 and 1909–10 he processed 858, 921 and 916 specimens respectively McAlpine (1910b). This involved hand-written responses to every enquiry, in his clear firm hand, as well as to all other correspondence.

In addition to all this, he was often asked to visit other states to help solve problems. He refers McAlpine (1910b) to such a case which occurred towards the end of 1908, when he was requested by the NSW Department of Agriculture to help resolve a problem in lemons and oranges, which when shipped or railed from Sydney to Melbourne often arrived in an unsaleable state. He travelled to Sydney, inspected fruit and sent consignments from Sydney to Melbourne, one set by rail and another by sea. These included samples with varying degrees of symptom expression, from none to extensive. He found that if growers carefully selected only fruit with no obvious black spots, it arrived in saleable condition. The black spot was caused by *Phoma citricarpa* (*Guignardia citricarpa*), a fungus apparently indigenous to Australia, which he had described McAlpine (1899a). He noted that affected fruit soon became rotten and was readily invaded by blue mould. He strongly advised that such fruit should not be exported overseas.

His own summary of his 20 years as Vegetable Pathologist McAlpine (1910c), which is reproduced in this Chap. 6, provides an excellent insight into the volume of work he managed, and the forward thinking behind his recommendations for providing adequate quarantine and expertise to protect Australian agriculture from the ravages of existing and imported plant diseases McAlpine.

5.3 Renowned Mycologist

Present day mycologists who have investigated species of earlier interest to McAlpine are generally struck by the thoroughness of his work. John Walker, now retired from the Biological Branch of the NSW Department of Agriculture, is an experienced taxonomic mycologist. In an appreciation of McAlpine's pioneering contribution, he makes the point that 'McAlpine the Mycologist' is intimately entwined with 'McAlpine the Plant Pathologist' and that his mycological work published between 1890 and 1911, was usually linked with his work on diseases caused by biotic agents, mostly fungi (Walker 2013).

One of his earliest publications in mycology that attracted wide attention was *Systematic Arrangement of Australian Fungi* McAlpine (1895b). This was a remarkable census of fungi found in Australia from 1770 to 1895 that included 2,290 taxa and 311 references (Walker 2013). It was set out as a ready reference in spread sheet form. McAlpine saw it as a fundamental tool for his work as Vegetable Pathologist. It was the first of six publications produced between 1895 and 1911 as handbooks. Those that followed describing diseases of citrus, stone fruits and potatoes, plus treatises on Australian rust fungi and Australian smuts. He wrote most of these in a manner easily understood by producers, students, advisors and instructors at all levels. They are considered individually in brief detail below (Sect. 5.6).

His astonishing contribution to mycology resulted in the description of 308 new species and 7 new genera of fungi from Australia, as well as 21 transfers of known species into other genera as new combinations. Five of his seven new genera, *Asteridiella, Laccocephalum, Leptosphaerulina, Schizotrichum* and *Uromycladium* are still accepted and in current use. *Asteridiella* now contains over 300 described species, and *Leptosphaerulina* contains over 25 species. Of the other two genera, *Amphichaeta* has been reduced to synonymy under the earlier name *Seimatosporium*, and the status of *Dimerium* McAlpine 1903, which predates *Dimerium* (Sacc. & P. Syd.) Sacc. & D. Sacc. 1905 remains to be fully resolved (Walker 2013).

Specimens that McAlpine studied are housed mainly in the herbarium (VPRI) at Agribio, the newly established facility of the Victorian Department of Environment and Primary Industries at La Trobe University, Bundoora. The Plant Pathology herbarium (VPRI), first established in McAlpine's Flinders Street workplace, was moved to Burnley in 1924, then in 1994 to Knoxfield and in 2013 to Bundoora. There are also some McAlpine specimens in the National Herbarium of Victoria (MEL). Some of these are duplicates of specimens in Herb. VPRI. Unfortunately samples of some species have been lost from the VPRI Herbarium, apparently leaving the MEL specimens as the only material of some species now remaining. For some of his new species, none of the specimens mentioned in the original descriptions, including type specimens, can be found in VPRI. While Mr. Ian Pascoe (2014, personal communication), who was the Victorian Department of Agriculture's Taxonomic Mycologist for many years, confirms John Walker's observations, he draws attention to the collection of McAlpine's microscope slides, which match the protologues of some species without type specimens. However, Pascoe IG (2014, personal communication) found that accession numbers on these slides match the accession numbers on specimens of other species with the same protologues. Examination of these specimens soon confirmed that the type specimen of species A was also that of species B, both of which were labelled species A! Pascoe also found that in many cases the fungi concerned were either rare or missing from specimens, and sometimes appeared to be incidental saprophytes, possibly produced in moist culture (some such slides were labelled 'in culture'), and were discarded. Also found in the VPRI Herbarium were a number of high quality 8 cm^2 lantern slides of diseased plant organs and also of sections of leaves containing fruiting bodies of various fungi. Unfortunately, accession numbers and names on some of these have been obliterated through the ravages of time.

McAlpine appears to have made relatively few collections himself (Walker 2013), which because of his love of field excursions seems strange. The most likely reason for this was the enormous escalation in his workload following his appointment as Vegetable Pathologist in 1890. Most of the collections he worked on were sent to him from Victoria, but some were sent from other States. McAlpine (1910f) listed the numbers of specimens he received. Regardless of who collected them, McAlpine put together a highly valuable well-documented collection, which is still used as a national resource, having been added to and kept up-to-date. The VPRI collection is importance in underpinning trade and market access for not only Victoria's, but also Australia's plant products and the importance of Dr. Ian Pascoe's contribution in his role as mycologist and VPRI curator for 30 years has been critical in keeping the collection intact and up to date. His efforts in maintaining this vital State and National asset cannot be overstated. The collection dates from McAlpine's appointment to the present day. It is in constant use as the reference point for diagnostic services, and matters relating to quarantine and trade. The present curator of VPRI, Dr Jackie Edwards, attests to it constant use and unique value in underpinning Australian trade and market access. Its survival owes much to Ian Pascoe's foresight and persistence.

His assistants, G.H. Robinson [1900–1908] and C.C. Brittlebank [1908–1911] added many hundreds of collections to the herbarium and prepared many hundreds of photographs of spores taken under the microscope to illustrate his publications, especially those of rusts and smuts. J.G.O. Tepper sent specimens from South Australia, F.M. Bailey from Queensland, and N.A. Cobb a smaller number from New South Wales. McAlpine ensured that a record of all the specimens he received was entered in hand-written registers, which are still kept in VPRI. According to Pascoe, while entries in the earlier volumes of the register were made in McAlpine's handwriting, those in later volumes of the register are in a different hand, possibly Robinson's. As time passed fewer entries were made by McAlpine, most being made by Robinson and later Brittlebank. Similarly, most microscope slides were labelled in Robinson's hand, although Brittlebank relabelled many of them with the original labels left beneath. In some cases transcription errors on labels applied by Brittlebank were evident once the original Robinson labels beneath them were revealed, and through cross-checking with the corresponding entries in the register. Entries in the register are still a valuable source of information about specimens, often clarifying details that are missing or confused on the specimen packet labels (Walker 2013). Unfortunately, Pascoe and Walker each found that many of the specimens McAlpine studied are now of variable quality. Walker found that with some exceptions McAlpine's collections of rust and smut specimens are large and in good condition, with the fungus as described still visible. On the other hand, some specimens of microfungi now consist of only one leaf without any sign of the fungus he described. Examples are some of the new species he described on citrus and wheat leaves. The precise identity of some of these will never be known (Walker 2013). A further complication with some of the correspondence between specimens and accession numbers arose from the situation where sometimes McAlpine found two or more new species on a single leaf, which then became the type material for two or more species. Unfortunately, although McAlpine and Robinson applied superscripts to the accession number to distinguish the slides of the separate species, most of these were lost during the relabelling by Brittlebank.

Both McAlpine and Brittlebank were conversant with shorthand, which they used readily (Wedge 1984). It is likely, too, that Robinson used it and that McAlpine made notes on the examination of specimens in shorthand for his assistants to transcribe into longhand in registers and onto labels. This may sometimes have resulted in errors.

5.4 McAlpine's Training in Mycology

Where and from whom McAlpine received his training in mycology is a question that has been asked. Walker (2013) observed that right from the start of his career in Australia he had a good knowledge of taxonomic mycology. His first mycological paper on the life cycle of a rust fungus (McAlpine 1891) was published within a year of his appointment as vegetable pathologist. Clues, however, may be found in the teaching atlases, where in the *Biological Atlas* published with his brother, there are detailed labelled drawings of micro- and macro-fungi (Fig. 5.4). No doubt he learned something about them in lectures by Thiselton-Dyer and possibly also Huxley. The available evidence, however, suggests that he was largely self-taught. He translated the atlas produced by the Dodel-Ports (1878–1883), which contained information on fungi (McAlpine 1880–1883). Importantly, he had access to de Bary's German text of 1866, the English translation (de Bary 1866) appeared in 1887, 4 years after McAlpine published Part 2 of the *Botanical Atlas* in which he used some of de Bary's illustrations. For example, Fig. 8 of Plate XXII in the Atlas was the same as Fig. 164b in de Bary's text. Similarly, his illustrations of the stages in the life cycle of *Puccinia graminis* in Plate XVI of the Atlas was composed from illustrations of various stages taken from publications by Dodel-Port, de Bary and Tulasne.

McAlpine was known to be an avid reader, and read widely. His students in Scotland (Anon 1882) and Australia (Anon 1932) greatly admired his scholarship and his enormous knowledge of the literature. Eulogies by past pharmacy students (Anon 1932) were full of praise for his two lectures devoted to the life cycle and other aspects of *Claviceps purpurea*. It is quite probable that he gave similar lectures to pharmacy students in Edinburgh prior to 1884. Having translated the Dodel-Ports' text he would have been familiar with its mycological content, and it is highly likely that once he obtained de Bary's text he would have studied it closely. Furthermore, it is evident that through Thiselton-Dyer, McAlpine went on visits to Kew Gardens. McAlpine maintained contact with Thisleton-Dyer (1904) and it is probable that through their friendship, he met Mordecai Cooke at Kew prior to his departure for Australia. Cooke began studying Australian fungi in or about 1880, and would have been interested to meet McAlpine when he visited Kew to collect a testimonial from the Director, Sir Joseph Hooker (1884). There seems little doubt that McAlpine knowledgeable of fungi before he arrived in Australia was well grounded, but his interests were broad and his focus on the fungi did not sharpen until the late 1880s when he experienced firsthand the destruction of which they were capable.

Baron Sir Ferdinand von Müller possibly fuelled Cooke's interest in Australian fungi by sending specimens to Kew (English 1987). Cooke published several descriptions of these fungi in the English periodical *Grevillea*. Then at the suggestion of von Müller, and with an invitation from Australian Colonial Governments,

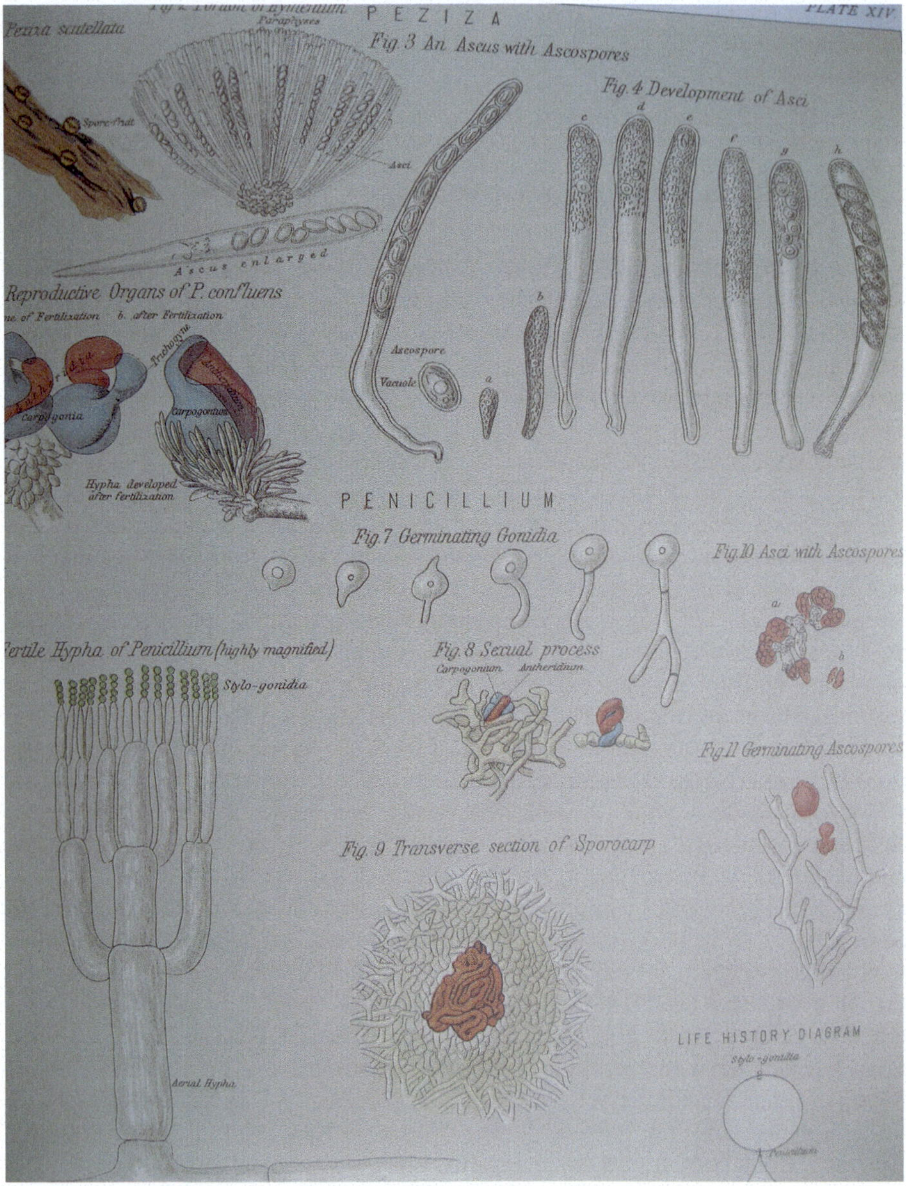

Fig. 5.4 Illustrations of *Peziza* and *Penicillium*. *Botanical Atlas* Vol. 2, Plate XIV (1883) (Courtesy of the Royal Botanic Gardens Library, Melbourne)

Cooke (1892) prepared a *Handbook of Australian Fungi*. In addition, von Müller was influential in McAlpine's involvement in the rust-in-wheat investigation and his appointment as Vegetable Pathologist. These appointments provided excellent opportunities for McAlpine to further develop his knowledge of mycology. It seems likely that McAlpine had begun his compilation of *Systematic Arrangement of Australian Fungi* (McAlpine 1891) soon after 1890, possibly with Cooke's knowledge and encouragement. Whereas Cooke began his study in about 1880 and took 12 years, McAlpine took 5 years. Many of Cooke's specimens were sent to him on loan (English 1987; Walker 2013).

5.5 Hallmarks of Rigour and Accuracy

McAlpine was a consummate scholar and student of natural history generally. This was recognised by his peers and his students in both Edinburgh and Melbourne. Once his interest was aroused he set out to learn as much about a subject as was available. The fact that he read widely and corresponded regularly with his peers demonstrates a great capacity for self-education in whatever took his interest. John Walker (2013) observed that as well as keeping in touch with his interstate colleagues, he kept in touch with many of the leading mycologists of the time, including G.E. Massee and C.B. Plowright in England, P. Dietel, P.C. Henning F. Ludwig and P. Sydow in Germany, P.A Saccardo in Italy, and G.P. Clinton in the United States of America. McAlpine also corresponded with his American friend Professor H.H. Whetzel, a plant pathologist with a keen interest in the taxonomy of certain groups of plant parasitic fungi such as the Sclerotiniaceae. McAlpine's search for information had a long reach.

Langdon and Fullerton (1977) noted the detailed accuracy of McAlpine's descriptions of fungi when they described a new genus of smuts based on *Macalpinomyces eriachnes* (Thüm.) Langdon & Full. The type material, collected in northern Australia by von Müller in 1855, was sent overseas where two mycologists studied it independently. Neither described the large, smooth, thick-walled cells which McAlpine noted were abundant among the spore masses. Von Müller's contacts ascribed the smut to different genera. Thümen (1878) named it *Sorosporium eriachnes*, while Cooke (1879) called it *Ustilago australis*. Langdon and Fullerton (1977) found that the ontogeny of the smut was distinct from that of smuts in either of the above genera, and confirmed McAlpine's description of the species. Dr Ray Langdon (2013) remarked that when the breadth and depth of McAlpine's activities are considered, he stands out as a great all-rounder.

John Walker (2013) observed that in his mycological papers, McAlpine usually gave an accurate description of all the fungi mentioned, whether they were new or well-known species. These descriptions of known species are valuable as they are based on Australian collections and in some cases supplement the characteristics for the species known previously only from the Northern Hemisphere. His descriptions of new species are detailed and complete, and often accompanied by line drawings of spores or photomicrographs. Often there is an extended discussion of the taxonomic relationships of the fungus, and the characteristics of the disease with which it was associated. The collections studied in preparing his descriptions are listed briefly, but often he does not clearly indicate a type specimen. This applies in the case of species where several specimens were listed, and has required extensive lectotypification by later workers. In some cases it is possible to determine his intention about typification either from the specimen label, or from the entry in the accessions register, where the notation '*sp. nov*.' indicates his chosen type specimen. This failure of his usual rigour may point to a lack of formal training as a taxonomist.

Walker (2013) has also observed that in all his writings, McAlpine adopted a narrow species concept. He described morphologically identical or very similar fungi on related hosts as different species. This was a common philosophy at the time and well illustrated by his treatment of the demicyclic rusts that occur on various genera of the *Asteraceae* in Australia. The rust that infects species of *Bellis, Brachycome, Calendula, Calotis, Lagenophora, Podolepis, Senecio* and *Vittadinia* amongst others, was considered by him, to be eight distinct species, *Puccinia distincta* McAlpine, *P. brachycomes* McAlpine, *P. calendulae* McAlpine, *P. calotidis* McAlpine, *P. lagenophorae* Cooke, *P. podolepidis* McAlpine, *P. tasmanica* Dietel and, P. *vittadiniae* McAlpine, respectively. This rust produces aecial aecia, uredinial aecia and teliospores on all hosts, the spore states are similar in morphology, limited cross infection occurs between hosts, and recent DNA studies indicate that only one taxon is involved for which the earliest binomial *P. lagenophorae* Cooke is available (Scholle et al. 2011; Wilson et al. 1965). This 'splitter' philosophy was evident in his papers describing fungi from various taxonomic groups, as it was common in the publications of his contemporaries. McAlpine was obviously passionate about microfungi. His prolific output in just 21 years, and his descriptions of so many fungi parasitic on cultivated plants and native flora, laid the foundation not only of plant pathology, but also of taxonomic mycology, in Australia. His publications soon became known far beyond our shores and established his international reputation. Walker (2013) concluded by noting that 'McAlpine has been called the father of plant pathology in Australia' but he is also 'Australia's Pioneer Mycologist'.

5.6 Six Australian Books

An appreciation of the depth and breadth of his knowledge and thinking emerges from his books, especially the prefaces but also the main texts. In various instances he foreshadows disease control programs such as pathogen-free schemes and the use of heat therapy to eradicate inoculum. A glimpse of this may be gained here by considering some of the commentary in his various Australian publications.

Systematic Arrangement of Australian Fungi (1895) In the foreword to this highly praised publication, McAlpine states that when he saw the magnitude of the task before him as Vegetable Pathologist he recognised a need to have as much possible basic information about each of the fungi recorded in Australia readily available. He built on M.C. Cooke's *Handbook of Australian Fungi* (Cooke 1892). He drew information from 310 papers – his own and many published by Australian authors such as Bailey and Bancroft as well as European authors including Berkley and the Tulasne brothers. The information for each species was tabulated across a double page in six columns under the headings: Name [Latin binomial], Authority, English name, Habitat, Occurrence, and General Characteristics. The 236 pages provided very useful information about mostly plant pathogens known by the mid-1890s.

Fungus Diseases of Citrus Trees in Australia and Their Treatment (1899) This is a book of 132 pages of text, twelve plates of excellent colour illustrations of symptoms by Brittlebank (Fig. 5.4) and a further 19 plates of 186 meticulously drawn figures of 69 pathogens, many of which he found only on fruit imported from other states, Italy or other overseas sources. The cost to Victoria of imported citrus fruits from Italy in 1897–98 was £62,498, the equivalent of $18,750,000 today. At the time, the Victorian citrus industry was in its infancy. Of the 401 acres of orange groves in all Victoria, 369 [92 %] were in the Mildura district where Alfred Deacon[A1] had encouraged the Chaffey Brothers to establish their irrigation scheme during the mid to late 1880s. Similarly, of the 552 acres of lemons grown in Victoria 81 % grew in the Mildura District. Many of the diseases described in this book did not occur in Victorian citrus groves, but McAlpine found them on fruit imported from elsewhere and was familiar with them. Of the 82 species affecting citrus fruit and trees he described 47 as new. In a brief concluding general remarks section of the book, he reminded growers that there is more to producing a crop than disease control. He directed attention to considerations of soil type, appropriate soil fertilisation and the need for liming to reduce the occurrence of certain diseases in some situations, and recommended the use of sprays to control other pests and diseases (Fig. 5.5).

Fig. 5.5 Charles Clifton Brittlebank (1862–1945) (Permission, Department of Environment and Primary Industries, Victoria)

Fungus Diseases of Stone-Fruit Trees in Australia & Their Treatment (1902) This book of 134 pages contained 10 water colour illustrations, some by C.C. Brittlebank and others, as well as 327 figures, produced by McAlpine (1902). In the introduction McAlpine drew attention to the common view among many growers that diseases resulted from the effects of sunscald, frost, flood and drought, and that there was little they could do about them other than wait for better seasons. He then quoted letters from proactive orchardists who had adopted spray programs. He also gave statistics on the increase in volumes of saleable fruit obtained through spray control programs. Thus he used his book not only to provide basic information about the pathogens and how they caused diseases, and instructions on how to prepare spray mixtures and apply them, but also as an extension tool to help educate growers about the possibilities and benefits of disease control programs.

The Rusts of Australia – Their Structure, Nature and Classification (1906) Of all his publications, this book (McAlpine 1906a) possibly had the greatest impact on his readership. This was in part because it dealt with the group of diseases seen as the greatest threat to food production in the Western world. The book was also highly valued for the scholarship it entailed and as the authoritative English language text and reference book. It is evident from the preface to this book that McAlpine's decision to gather information on rusts was encouraged by colleagues involved in the Rust-in-Wheat Conferences who recognised that numerous rusts had the potential to devastate a range of agricultural industries other than those based on wheat and other cereals. This has proved to be the case and McAlpine's belief that the more detailed our understanding of the life cycles, states and stages within them, their aetiology and ecology, the stronger our knowledge base from which to fight them (Fig. 5.6).

It is evident from what he wrote in his Chap. 19, that much of the poor advice being given at the time concerning rust mitigation relied on a meagre understanding of the disease. McAlpine's own conclusions were that the most reliable means of controlling rusts were:

1. To produce wheats suited to Australian conditions by crossing, as Mr Farrer was then doing.
2. To select and carefully cultivate the most rust-resisting plants from these and other wheats having the necessary hardy, prolific, grain-holding and milling qualities, keeping up the strain [variety] and constantly renewing strains [variety] and keeping up their constitutional vigour.
3. To cultivate early maturing varieties and sow them early, combined with the best agricultural methods such as clean cultivation, judicious rotation, suitable (phosphatic) manuring and fallow, to insure a good crop.

McAlpine's understanding that once wheats with desirable agronomic characteristics, milling and baking qualities and rust resistance were found it was not a permanent fix! He foreshadowed a continuous need to incorporate into such wheats new combinations of genes for resistance to rust to keep pace with the ability of rust fungi to produce new genes for virulence. This has proven to be the case, with plant breeding establishments in many countries established for just this purpose, not just for rust in wheat but for other crops and their various diseases.

Fig. 5.6 Life Cycle of *Puccinia graminis*. Botanical Atlas Vol. 2, Plate XVI (1883). Composed from Dodel-Port, De Bary and Tulasne (Courtesy of the Royal Botanic Gardens Library, Melbourne)

In his 1892 *Handbook of Australian Fungi*, M.C. Cooke included 72 species of rusts described and named from Australian specimens collected by various early explorer-collectors and returned to Europe and the United Kingdom. In *Rusts of Australia*, McAlpine expanded this to 161 species, describing one new genus,

Uromycladium, 56 new species and changing the names of six others. In addition McAlpine provided detailed descriptions of rust life cycles, and their complexities as well as providing highly accurate descriptions of infection processes and explanations of how such plant pathogens robbed their hosts of sustenance. Contemporary researchers have confirmed most of what he described and much of his speculation regarding how rust fungi affect their hosts. The book contains 55 plates including 366 figures. The first plate, the Frontispiece, shows the destruction of a grove of golden wattle, *Acacia pycnantha*, by *Uromycladium tepperianum* (Sacc.) McAlp. At that time wattle bark was in high demand for tanning leather, so that groves were planted to increase the efficiency of collecting wattle bark. Because of the rust, however, the practice was abandoned. Ten plates show watercolour illustrations of symptoms on various hosts executed by Brittlebank, and the remainder are photographs and photomicrographs of rust spores and related subjects taken by McAlpine's assistant of the day, G.H. Robinson, as well as figures drawn by McAlpine himself. The book also contains a comprehensive glossary and host index. The contents represent an enormous amount of detailed work of the highest calibre, and provide valuable information and intellectual insights.

Rusts of Australia propelled McAlpine on to the world stage and focused international attention on him. He received high praise from Professor J.C. Arthur in a review published in the *Journal of Mycology* of March 1906. Arthur was regarded at the time as the expert on rusts. Professor C.E. Bessey's enthusiastic review of the book appeared in *Science* in January 1907. In a review published in *Revue Général des Sciences Pures et Appliqués* in August 1907, Professor P.A. Saccardo referred to McAlpine's *Rusts of Australia* as 'a magnificent volume, a very important work'. Other reviews appeared. One in the *Journal of Botany* in August 1906, another in *The Victorian Naturalist,* also in August, and a salutary one appeared in the *London Times Weekly Supplement* of 24th May 1907. The praise for the volume increased public awareness of McAlpine's prominence as a mycologist and plant pathologist. From the early 1890s he assisted William Farrer by testing his wheat selections and crosses for resistance to rust when grown in Victoria, as Cobb did in New South Wales, and Angus claimed to have done in South Australia (Daniels 1979). McAlpine also tested other cereals for rust susceptibility, including barley (Waterhouse 1922). The book cost £1/10/0 [$3].

Smuts of Australia -Their Structure, Life History, Treatment and Classification (1910) While the importance of McAlpine's research into the life cycles of rusts and his involvement in the Rust-in-Wheat Conferences is widely understood, his efforts to stem the dramatic impacts of cereal smuts and bunts (McAlpine 1910d) seem to have attracted less attention, although they were equally important. At the time, stories were told of how some crops were so heavily infected with bunt that during harvesting, fields were covered in blinding black clouds of spores, which posed a danger. On at least one occasion, when a farm hand struck a match to light a cigarette, it was claimed that the spore-cloud exploded and fire enveloped the harvester, injured the farmhand and burned out hundreds of acres of surrounding crop. During his first decade, a great deal of time was spent experimenting with methods for bringing such epidemics and catastrophes under control.

Unlike the cereal rusts which share a common infection pattern, the smuts are carried over from season to season by different means and infect at different stages of their hosts' development. This means that there is no common approach to controlling them. McAlpine explained that among the eight cereal smuts causing problems in his day, there were three infection patterns requiring three different approaches to control. Two smuts, loose smut of wheat and naked smut of barley, were 'flower-infecting smuts'. Smutted plants released their spores at the time healthy plants were developing grain and the smut invaded the developing seed without damaging it and without producing symptoms. The following season, the infected seeds germinated normally but the smut fungus replaced the developing inflorescence and released spores in time for them to invade the developing grain in adjacent healthy plants. The control recommended here was to subject seed-wheat to hot water treatment, which when carried out under well-controlled conditions was successful. Three smuts were of the second infection type, and known as 'seedling infecting smuts'. Seed wheat which had come from smutted or bunted crops was usually thoroughly dusted with spores during thrashing. The spores remained attached to the seed and germinated with it, infecting the seedling. Steeping the seed in fungicidal solutions or dusting with fungicides was highly effective. The third type was 'seedling and shoot infecting smuts', of which flag smut is an example. Spores survived in stubble and soil and infected seedlings directly. Spores produced on infected grasses or infected wheat early in the season could be carried by wind to infect seedlings of later crops in the vicinity. Thus they were able to infect seedlings or young shoots. Crop rotation was advocated but later highly stable resistance genes proved the best control.

This publication was fully illustrated with 312 high quality photographs and photomicrographs by G.H. Robinson and C.C. Brittlebank, contained detailed information on 67 species of smuts, 27 of which were named by McAlpine, as well as detailed explanations of their biology and pathology. A host index was included together with a philosophical discussion on the possible uses of immunity and resistance. It sold for £1/5/0 [$2–50].

Fungus Diseases of the Potato in Australia and their Treatment (1911) McAlpine dated the Preface to this book 6 weeks after his appointment as leader of the national investigation into bitter pit. During the months leading up to the appointment, he had expressed his wish to complete the work and make it available to the industry. Various comments he made in his correspondence during this time give a sense of urgency in his wish to have the book published and available to the public. This was because of concern over the rapid spread of Irish Blight throughout Australian potato growing districts during the previous 2 years. It was a comprehensive handbook, useful to farmers, advisers, teachers and students. In its first 202 pages it dealt with the diseases of potatoes in Australia, with a further 13 pages of appendices providing useful information. The first appendix relating to weather at first appears curious because it deals with only the first 3 months of 1911. It was, however, because he suspected there was a connection between weather conditions and rate of spread.

Tryon (Fig. 5.7) (1909) had recorded the first outbreak of Irish Blight in Australia on potato and tomato in Queensland on 19th May 1909 (Tryon 1909). By 31st August that year it had been found in all states (McAlpine 1909). It was found at Fish Creek

Fig. 5.7 Henry Tryon (1856–1943) c.1920 (Courtesy, Royal Botanic Gardens Library, Melbourne)

in Victoria on 23rd July 1909 by which time Government Officers in all states were deeply concerned about its rapid spread and how to contain it. They were aware of the impact of the Irish famine of 1845 (Fig. 4.1) and the annual cost to the North American potato industry, which was estimated at £7,500,000 [Aust. $2,000,000,000 today] and the outbreak in New Zealand in 1905, which had cost growers £200,000 [Aust.$3,000,000 today]. Thus, when McAlpine noted the increasing rate of spread in Victoria, from six collections between July and November 1909 followed by 71 specimens between June and December 1910 and then 152 specimens found during the first 3 months of 1911, which were wetter and more humid than average (McAlpine 1909), he was alarmed. His concern was the possible reason for the first 55 pages of the 215-page book being devoted to Irish Blight. Even so, the other ten groups of diseases were well covered, detailed and carefully illustrated. A second appendix set out Commonwealth regulations and a proclamation on Quarantine restrictions. A third appendix related to Victorian legislation. Two further appendices 'A' and 'B' were added. The first, written by William Laidlaw, introduced readers to the biology of eelworms through a well-illustrated account of the biology and control of the golden cyst nematode. The second provided succinct accounts of the main insect pests of potato, their biology and control, written and illustrated by Charles French Jnr., then Acting Government Entomologist. This publication also sold for £1/5/0 and was republished some years later as *Handbook of Potato Diseases in Australia*, selling for the same price.

McAlpine published other work on Irish blight. Its occurrence on tomatoes in New South Wales (McAlpine1910e) and had tested potato varieties for resistance to Irish blight (McAlpine 1910f).

5.7 Extra Mural Activities

McAlpine loved field excursions and finding fungi new to science. He was an active member of the *Victorian Field Naturalists Club* and published 17 papers on Victorian fungi in their journal between 1893 and 1910, describing new species of macro- and micro-fungi. Eventually his workload increased to a level where he was unable to find time for collection other than during excursions with his students, so that he became increasingly reliant on collectors and specimens sent for identification. He was also a corresponding member of the *Linnean Society of New South Wales* and published 19 papers in their proceedings between 1895 and 1911. He attended meetings of the Royal Society of Victoria and presented talks on his findings and ponderings. For example, he gave one talk on the problem of determining whether myxomycetes were plants or animals (McAlpine 1899b), ending it with a saying often used by his old teacher and friend Thomas Huxley that 'the next best thing to being certain of what is certain, is being uncertain about what is uncertain!'

He wrote three articles in German (McAlpine 1892, 1898a, b), which he published in *Zeitschrift für Pflanzenkrankheiten,* for which he was one of the 44 members of the *International Phytopathological Committee* responsible for maintaining it. He also assisted others scientists to solve problems which were not necessarily plant pathological in nature. For example, in 1893 he published the results of research into the sudden death of cattle at Pascoe Vale near Melbourne. This was his only known foray into forensic pathology. He assisted P.W. Farmer MB (McAlpine and Farmer 1893) in proving the cause of deaths in cattle was ingestion of an introduced South African weed, *Homeria collina* Vent. var. *minutae*. They demonstrated the weed's lethal effects on rabbits and cattle in controlled experiments.

Curtis Gates Lloyd was apprenticed to a pharmacist in 1867. By 1885 he and his brother Uri launched a pioneering book *Drugs and Medicines in North America* which was popular in the fields of medicine and botany. Two years later, C.G. Lloyd became interested in mycology partly for aesthetic reasons and partly for their pharmaceutical potential. At this time McAlpine (1885, 1886) also published papers of pharmaceutical interest. By 1907 the Lloyds were men of independent means with offices in America, England and France, and in 1908 C.G. Lloyd began publishing *Mycological Notes* for which the price of a subscription was 'collecting and sending him specimens of basidiomycetes other than fleshy agarics'! From time to time Lloyd circulated requests for specimens of polypores and phalloids as well as notes on how to photograph fungi and how to prepare specimens for posting. McAlpine and Lloyd each had common interests in pharmacy and botany which they shared through correspondence. Lloyd published two of McAlpine's notes on Australian phalloids; *Anthurus aseroeformis* (McAlpine 1908) and *Jansia truncata* (McAlpine 1910a) and McAlpine sent Lloyd specimens of interest including phalloid species and other publications such as those regarding *Polyporus mylittae* (McAlpine 1893a, 1904d) and possibly specimens of luminous species (McAlpine 1900e).

McAlpine was also interested in entomogenous fungi (McAlpine and Hills 1894), such as species found on beetles (McAlpine 1896), locusts (McAlpine

1900a), codlin moth (McAlpine 1904a) and scale insects (McAlpine 1904b). He also published one nematode paper, about *Tylenchus devestatrix* on onion (McAlpine 1895a), and one on a bacterial disease of mulberry (McAlpine 1898a). The extent of his collecting and breadth of his interests is merely touched on by what is recounted here. The size of his personal collection of fungi numbering approximately 10,000 specimens may be a better gauge.

McAlpine's observations on methods of plant disease control demonstrate that he thought ahead of the times and saw needs for control practices which sometimes only became common practice half a century or more later. In his book on potato diseases when dealing with the problem of Irish Blight he emphasised the importance of using pathogen-free planting stock. At the same time, possibly in part because of observations made by Grasby (McAlpine 1911) that Irish Blight was rare in areas of Western Australia subject to hot growing conditions, and partly because of his experiments on controlling embryo-infecting smuts through hot water treatment, he experimented with heat treatments of potato tubers as a means of providing pathogen free planting stock (McAlpine 1911). He achieved some success, but it was expensive and needed very careful temperature control. Because there were other sources of mostly airborne inoculum including the inevitable failure of all growers to use pathogen-free planting stock, and because of the rapid, extensive spread of air-borne inoculum, other methods were essential and heat treatment was abandoned.

Although he was appointed on a part-time basis as a consultant, McAlpine did more work during those 20 years than many do in a lifetime. Not only did he need to diagnose many diseases previously unknown to science, but he also confirmed their causes and devised practical control measures. His other commitments included regular lectures, demonstration sessions, and field excursions for the Pharmacy College and the Burnley School of Horticulture, conducting a personal extension service to farmers, reading widely, conducting a whole orchestra of research projects, and publishing his findings. Prior to arriving in Australia, between 1877 and 1883 he had published eight articles and atlases and had three more in press. From 1885 to 1889 he produced 12 more. After his appointment in 1890, however, he exploded into print! During the first decade he published 80 bulletins, reports and articles and during the second decade 107. From 1910 to 1916, when he was heavily preoccupied with the bitter pit investigation, he wrote 20 reports and articles. At first glance (McAlpine 1910a) it seems that many of his listed publications were single page bulletins. However, it was common practice among authors of that period to note only the number of the first page of an article and provide only the minimum information necessary for the reader to find what was required.

McAlpine was widely read and maintained a worldwide network of correspondents, keeping himself fully informed and up-to-date. His contacts were universal, and he maintained an active correspondence, always seeking new information or providing assistance to those who appealed to him. For example in October 1900 he received a letter from Erwin F. Smith (1900) then Pathologist-in-Charge of the Plant Pathology Laboratory and Director of Vegetable Physiology and Pathology at the US Department of Agriculture, Washington DC., who provided the identity of *Pseudomonas campestris* and explained that there was no book then available on

Fig. 5.8 British Association visitors on a Bay-side excursion with McAlpine in 1914 (Permission of the State Library of Victoria, Melbourne. La Trobe Australian Manuscripts Collection MS 12772, Box 3555/3)

bacterial diseases. McAlpine also spent a great deal of time instructing farmers on the need and procedures for controlling diseases. Much of this advice was based on personal experience or knowledge of control treatments devised elsewhere. He gave lectures to agricultural advisors at Burnley, pharmacy students in Parkville and other groups at various venues.

Above all, McAlpine was a peoples' person. He was a loving family man. He thoroughly enjoyed his students who rewarded him with their attention and friendship (Anon 1882). He enjoyed working with those on the land who increasingly sought his help for which they were grateful. He was greatly respected by his work colleagues and peers, and enjoyed meeting others interested in the sciences. He extended his vast network of contacts and his reach for information through the extension of his hospitality (Fig. 5.8).

McAlpine attended scientific gatherings, participated in meetings and was respected for his knowledge and insight. He delivered the first monthly public lecture given by specialists at the School of Horticulture, Burnley in 1891 (Anon 1893). He regularly attended conferences run by the Australian Association for the Advancement of Science and spoke at them. He spoke at the Adelaide conference on botanical nomenclature as applied to fungi (McAlpine 1893b) and he spoke again at the 1897 Sydney conference (McAlpine 1898c) giving an account of Australian fungi. He also attended other conferences held in Australia such as that of the 1914 British Association for the Advancement of Science held in Melbourne (Fig. 18.2). His talks usually contained philosophical themes, humour and a touch of lightheartedness, such as in a talk delivered to the Victorian Field Naturalists Club on *The Romance of Plant Pathology* (McAlpine 1910g).

References

Ainsworth GC (1981) Introduction to the history of plant pathology. Cambridge University Press, Cambridge, 315 pp

Anon (1882) Letter [undated] signed by 30 students at the Watt Institute and School of Arts supporting McAlpine's application for a post in Otago

Anon (1890) Centennial international exhibition Melbourne 1888–1889. Sands & McDougall, Publishers, Melbourne

Anon (1893) Monthly lectures delivered at school of Horticulture, Burnley, by various specialists during 1892–1893. Government Printer Melbourne No. 826, for Department of Agriculture, Victoria, on behalf of School of Horticulture Established May 1891

Anon (1932) Death of Daniel McAlpine. Australas J Pharm, 13:933, 1003–1005, 1037, 1096

Argus Newspaper (1891) A report [30-7-'91] of the death of Mr. Smith Ellis while attending his rust control demonstration in the Melbourne Botanic Gardens

Colac Herald (1890) Reporting [13-5-'90] that Mr. Smith Ellis's claimed the £10,000 reward for his treatment to control wheat rust, and its worth being doubted by Professor W Brown and Mr. AN Pearson

Cooke MC (1879) Undescribed fungi in the Kew Herbarium. Grevillea 8:34–35

Cooke MC (1892) Handbook of Australian fungi. Williams & Norgate, London for the Departments of Agriculture in Melbourne, Brisbane, Sydney Adelaide, Hobarton

Daniels JG (1979) Angus, William (1871–1965). Aust Dict Biog 7:74–75

De Bary A (1887) Comparative morphology and biology of the fungi mycetozoa and bacteria. (Authorised English Translation). Clarendon Press, Oxford, 525 pp

Dodel-Port AC (1878–1883) Erläuternder Text aus Anatomisch-physiologischen Atlas der Botanik für Hoch- und Mittleschulen. University of Zurich

English MP (1987) Mordecai Cubitt Cooke, Victorian naturalist, mycologist, teacher & eccentric. Biopress Ltd, Printed in Great Britain by Henry Ling Ltd, at Dorset Press, Dorset ISBN 0–948 737-02-6

Fish S (1976) Daniel McAlpine – a pioneer plant pathologist in Australia. Aust Plant Pathol Newsl 5(1):11–14

Gippsland Times (1890) Report [30-4-1890] that the Victorian Government Chemist, Mr. A.N. Pearson's doubted Mr. Smith Ellis's secret seed dressing would control wheat rust, while the Minister for Agriculture's supported testing it

Henzell T (2007) Australian agriculture- its history and challenges. CSIRO Publishing, Melbourne, 308 pp. ISBN 9780643993426

Hooker SJ (1884) Letter [26-5-'80] written to Mr. Abbott, curator of the Hobart Botanical Garden, with a testimonial to McAlpine's abilities

Kilmore Free Press (1890) An article [17-4-'90], *Rust in Wheat Cure*, dealt with the offer of a £10,000 reward for Mr. Smith Ellis should his alleged control treatment for wheat rust be successful, and included interviews with Sir Henry Parkes, Premier of New South Wales, and the Premier of Queensland

Langdon RFN (2013) Letter (15-7-'13) to the author noting his experience of the breadth and quality of McAlpine's scholarship

Langdon RFN, Fullerton RA (1977) *Macalpinomyces*, A new genus of smut fungi. Trans B Mycol Soc 68(1):27–30

McAlpine D (1880–1883) Translated and edited text of Dodel-Ports'. In: Explanatory text from the anatomical and physiological atlas of botany. W and AK Johnston, Edinburgh/London, 376 pp

McAlpine D (1885) Modern botany in its relation to pharmacy. Aust J Pharm 31:7–9

McAlpine D (1889) Chairman's report on vegetable products and cultivated vegetable products. Centennial International Exhibition, Melbourne, pp 772–778

McAlpine D (1891) The life-history of the rust of wheat. Bulletin No.14, Department of Agriculture, Victoria 8 pp and Plates I and II

McAlpine D (1892) Ueber die Verwendung geschrumpfter Körner von rostigem Weizen als Saatgut [Regarding the use of grain shrunken through the affect of rusted wheat]. Zeitschrift für Planzenkrankheiten 2:193

McAlpine D (1893a) Native bread (*Polyporus mylittae* C & M). Aust J Pharm 8:291

McAlpine D (1893b) Botanical nomenclature with special reference to the fungi. Aust Assoc Adv Sci Adelaide 5:414–420

McAlpine D (1895a) Report on onion disease (*Tylenchus devestatrix* Kuhn. In: Guide to growers No.18, Department of Agriculture Victoria. Government Printer, Melbourne

McApine D (1895b) Systematic arrangement of Australian fungi, together with host index and list of works on the subject. Department of Agriculture, Victoria. Government Printer, Melbourne

McAlpine D (1896) A fungus on a beetle. Vic Nat 13(4):56

McAlpine D (1898a) Bacteriankrankeit der Maulbeerbaume. [Bacterial Disease of Mulberry Trees]. Zeitschrift für Pflantzenkrankeiten 8:142

McAlpine D (1898b) Ueber die Anwendung von Fungiciden bei Weinstocken [Application of Fungicides to Grapevines]. Zeitschrift für Pflantzenkrankeiten 8:11

McAlpine D (1898c) A statistical account of Australian fungi up to the end of 1897. Aust Assoc Adv Sci Sydney 8:1–3

McAlpine D (1899a) Fungus diseases of citrus trees in Australia and their treatment. In: Department of Agriculture, Victoria. Government Printer, Melbourne

McAlpine D (1899b) Plant or animal – myxomycete or mycetozoon? Vic Nat 16:95–98

McAlpine D (1900a) The systematic position of the locust-fungus imported from the cape. Agric Gaz NSW 11:184

McAlpine D (1900b) Wheat experiments at port fairy and in the Mallee. Report of the Department of Agriculture, Victoria. Government Printer, Melbourne

McAlpine D (1900c) Report on the work of the pathologist's branch for the year ending 31st December 1899. Report of the Department of Agriculture, Victoria. Government Printer, Melbourne, pp 44–61

McAlpine D (1900d) Report for the year 1900. Department of Agriculture, Victoria. 36 pp, 11 Plates, 18 Figs. Government Printer, Melbourne

McAlpine D (1900e) Phosphorescent fungi in Australia. Proc Linnean Soc NSW 25

McAlpine D (1901) Fungous diseases of cabbage and cauliflower in Victoria and their treatment. Department of Agriculture, Victoria. Government Printer, Melbourne

McAlpine D (1902) Fungus diseases of stone-fruit. Department of Agriculture, Victoria. Government Printer, Melbourne

McAlpine D (1903) The hop industry in Victoria. J Dep Agric Vic 11(6):37–39

McAlpine D (1904a) A fungus parasite on codlin moth. J Dep Agric Vic 11(7):649–650

McAlpine D (1904b) Two new fungi parasitic on scale insects. J Dep Agric Vic 11(7):646–648

McAlpine D (1904c) Report of the vegetable pathologist [Including bitter pit] for 1903–1904. J Dep Agric Vic 12:1–9

McAlpine D (1904d) Native or blackfellow' bread. J Dep Agric Vic 12:13

McAlpine D (1906a) The rusts of Australia, their nature, structure and classification, 349 pp with 55 plates and 366 figures. Victorian Department of Agriculture. Government Printer, Melbourne

McAlpine D (1906b) A new hymenomycete – the so-called *Isaria fuciformis* Berk. Ann Mycologici 4:541–551

McAlpine D (1910a) List of scientific works, papers, bulletins, etc. 1877–1910, by D. McAlpine, Government vegetable pathologist. Department of Agriculture, Victoria, 12 pp, Government Printer, Melbourne

McAlpine D (1910b) Report of the vegetable pathologist for 1907–10, Department of Agriculture, Victoria. Government Printer, Melbourne, pp 44–61

McAlpine D (1910c) Twenty years of plant pathology in Australia. Report of the vegetable pathologist for 1907–1910. Department of Agriculture Victoria, Government Printer, Melbourne, pp 56–61

McAlpine D (1910d) The smuts of Australia, their structure, life history, treatment and classification, 312 illustrations. Department of Agriculture Victoria, Government Printer, Melbourne, 288 pp

McAlpine D (1910e) Irish blight in tomatoes. J Dep Agric Vic 8:48–49

McAlpine D (1910f) Testing potatoes varieties for Irish Blight. J Dep Agric Vic 8:358–359
McAlpine D (1910g) The romance of plant pathology. Vic Nat 27:127
McAlpine D (1911) Handbook of fungus diseases of the potato in Australia and their treatment, 161 figs. Department of Agriculture, Victoria. Government Printer, Melbourne, 215 pp
McAlpine D, Farmer PW (1893) Report on a poisonous species of *Homeria*, *Homeria collina* Vent. Var. *miniata*, causing death of cattle feeding on it. Proc R Soc Vic 5:396
McAlpine D, Hills WHS (1894) The entomogenous fungi of Victoria. Proc R Soc Vic 7:159–160
Scholle M, Lutz M, Wood AR, Hagedorn G, Mennicken M (2011) Taxonomy and phylogeny of *Puccinia lagenophorae*: a study using rDNA sequence data, morphological and host range features. Mycol Prog 10:175–187
Smith EF (1900) Letter sent by Erwin F. Smith to McAlpine confirming the identity of *Pseudomonas campestris*
Stalpers JA, Loerakker WM (1982) *Laetisaria* and *Limonomyces* species (Corticiaceae) causing pink disease in turf grasses. Can J Bot 60:529–537
Thisleton-Dyer WT (1904) A letter [15-12-'04] written to McAlpine
Thuemen F (1878) Symbolae ad florum mycologicam Australiae. Flora 36:440–444
Tryon H (1909) The potato disease. Queensland Agric J 23:118–127
Turner JC (1910) Report of the senior inspector, Fruit exports and imports. Report of the department of agriculture, Victoria for 1907–1910, cf p 221. Government Printer, Melbourne, pp 213–228
Walker J (2013) Letter of appraisal of McAlpine's mycological work sent to the Author
Waterhouse LW (1922) Letters written to the Victorian department of agriculture asking if they had the results of McAlpine's tests of barley lines for resistance to leaf rust
Wedge E (1984) Letters to the Author [4th to 7th May-1984], La Trobe Collection
Wilson IM, Walshaw DF, Walker J (1965) The new groundsel rust in Britain and its relationship to certain Australian rusts. Trans B Mycol Soc 48:501–511
Wrigley CW (1981) Farrer, William James (1845–1906). Aust Dict Biog 8:471–473

Chapter 6
Twenty Years of Plant Pathology in Australia 1890–1910

6.1 In McAlpine's Own Words

The concluding section (McAlpine 1910a) of McAlpine's contribution to *The Report of the Department of Agriculture for the Years 1907 to 1910* (McAlpine 1910b) is included here in full to provide an appreciation of the scope of the work he dealt with, the circumstances under which he achieved his outcomes and his views on what Australia needed at the time, in terms of plant pathologists, their training and their nation-wide employment. The contribution made by McAlpine to Australian plant pathology during these years is best described in his own words, as follow:

"As it is just twenty years since I was appointed Consulting Vegetable Pathologist to the Department of Agriculture in Victoria, the present seems a fitting opportunity to point out, in simple language, the aims and objectives of this Branch, the work it has accomplished, the means provided for investigation and research, how it has benefited the producer and thereby added to the wealth of the State, and finally what still remains to be done in the way of conquering disease.

The ideas prevalent on the subject [the cause of plant diseases in 1890] were so hazy and the view that the weather, in some mysterious way, was accountable for all the ills that plants are heir to, was so widespread and deeply rooted in the minds of politicians and practical growers, that the utility of or necessity for investigating or doctoring diseases in plants, was held to be beyond the range of practical politics. However, thanks to the enlightened and farseeing policy of the Hon. Alfred Deakin, when acting Minister of Agriculture, the position was created as above, and I was appointed on 12th May 1890, to attend to any diseases which might form subjects of inquiry. The importance and far-reaching nature of the study was not then realised, and I was only expected to devote a portion of my time to it, as the remuneration offered was too small, to allow of me giving up my other public engagements.

Not withstanding the drawback that the laboratory work had to be carried on in my private dwelling and the limitations as to the time that could be devoted to such a wide field, I think I may say without undue boasting, that Victoria leads the way among the States, in this important branch of study, and as a proof of this assertion it may be mentioned that the results obtained here are quoted in American, British and foreign journals and text-books, and that one of the leading authorities in the United States, Professor Arthur, in a review of the work on *The Rusts of Australia*, stated - "It is by much the best account now available in the English language".

D.G. Parbery, *Daniel McAlpine and The Bitter Pit*,
DOI 10.1007/978-3-319-09552-3_6

6.2 What Constitutes Plant Pathology

The name in rather unfortunate, whether we speak of Vegetable Pathology or Plant Pathology, for while the meaning of the term Pathology is generally understood as referring to disease, but when it is used in connexion with vegetable or plant, it conveys the impression to many that it is confined to vegetables grown in a garden or to plants which are not supposed to include those cultivated for profit, such as wheat, potatoes or fruit-trees.

Plant pathology really includes everything relating to whatever interferes with the healthy growth of a plant, but for practical purposes it is officially restricted to those diseases which are caused by insects and other animals on the one hand, and those caused by fungi and allied plant organisms on the other. To the practical grower it resolves itself into a study of insect pests and plant diseases, but it is evident that this view excludes some of those diseases, such as Bitter Pit in apples, which are due to neither and yet are among the most destructive with which growers have to contend.

The entomologists have revealed the life histories of injurious insects and the means of combating them, while my own work has been devoted to the fungi and their treatment and the insidious nature of the attack of these parasites, together with the hidden way in which they do their work of destruction, like the pestilence that walketh in darkness, prevents them impressing the popular imagination in the way a boring beetle, a burrowing grub or a codlin moth has done.

6.3 Its Aims and Objectives

Fungi are thus studied from the point of view of the practical grower, and the aim and objective in view is, so to trace their different stages and find out their leading characteristics, that modern skill and knowledge may circumvent them, and prevent the damage, which they naturally cause. To this end it is not only necessary to study them in the laboratory, where all the modern appliances are used to investigate their course of development and wonderful life history, but also in the field to trace their effects upon the plant and the symptoms they produce. And above all, fieldwork is necessary in order to test those methods of prevention in a practical way, and on a commercial scale, which the studies in the laboratory have suggested.

Fungi like other plants, originate from reproductive bodies, usually minute and known as spores, and it is generally against these bodies that preventive measures are directed, so as to nip the disease in the bud, for it is the peculiarity of most fungi, that if they once gain an entrance into the plant, the disease must run its course, since any treatment fatal to the fungus at this stage would also be injurious to the plant. There is an exception, however, in the case of Irish Blight, since the fungus inside the tuber may be destroyed by heat, while the potatoes thus treated grow freely when planted.

Numerous instances could be given where the destruction of the spores prevents the disease and this is most easily done at the time when they are particularly tender and susceptible to treatment; that is, when they are about to germinate and put forth a germ-tube which enters the plant. Black spot of the apple and pear, Stinking bunt of wheat, Head Smut of maize, and numerous other diseases can be prevented in this way.

There are two principle factors, which have rendered the study of these fungus diseases less attractive and less popular than the study of insect pests. The first is that the use of the microscope is indispensable and the second is that the true cause of the disease is not easy to discover, since there are often various moulds and bacteria, which are consequences not causes, and simply act as scavengers when decay is otherwise set up. It is here that both knowledge and experience are required, as there are many pitfalls for the unwary and ignorant.

The use of the microscope is not a difficult matter although in untrained hands it may be productive of more worry than profit, but I may say here that for the working farmer or orchardist it is not recommended, and he had better be guided by the results obtained by someone experienced in its use. However, it is worth recording as showing a desire to study these things at first-hand, that I have been asked to recommend a microscope for a prize-winner at an agricultural show, instead of the time honoured cup, and also for one of our most experienced and successful orchardists who wished to see for himself those wonderful spores of the Black Spot fungus which used to produce such loss and at one time threatened the apple-growing industry.

But while the grower is only practically interested in the treatment, and may well leave the microscope alone, I emphatically protest against the teachers of the farmer, who deal with fungus diseases, not having such knowledge as only the microscope can properly give. This is fundamental for a proper understanding of the subject, and I pity anyone who talks to farmers on the Irish Blight for instance, who had not himself seen it under the microscope at first hand. Drawings and textbooks are all very well, but they can never replace the object itself, for one who sets himself up as an expert. Hence it is that in this subject particularly there is much misleading and inaccurate information. The cause of the disease is not always easy to discover, and it is here that training and judgement come in. It is evident that this is the first point to settle, before there can be, as a rule, any reasonable method of treatment devised, and it is necessary to infect the plant with a particular fungus and produce the disease before certainty can be arrived at. In the case of the Irish blight fungus, spores were taken from a diseased potato, and the disease produced in the laboratory, both in healthy potatoes and tomatoes, while in the case of head smut of maize, the seed dusted with spores produced the disease when grown in the field and not in any other way.

6.4 Means Provided for Investigation and Research

At first, as already pointed out, all the work was done single-handed and in rooms provided by myself. Then as the work enormously increased, and after a lapse of 10 years, an Assistant was appointed in 1900. I was fortunate in having as my first

Assistant, Gerald H. Robinson, who was not only gold medallist at Longerenong Agriculturalist College, but had also a course of training in orchard work at Burnley Horticultural Gardens.

After 8 years' service, Mr. G.H. Robinson resigned, having accepted a much more lucrative appointment in the commercial world. Mr. C.C. Brittlebank, who was a practical microscopist, as well as a practical farmer, was appointed in his stead.

In the other States of the Commonwealth, there is only an official Plant Pathologist in New South Wales. The mother State was fortunate in having for a number of years such a distinguished plant pathologist as Dr. Cobb, who investigated the diseases of numerous cultivated plants, and in many instances devised measures for their prevention. In another sphere of labour as Crop Technologist in charge of Agricultural Technology, Bureau of Plant Industry, Washington, he still remembers his congenial work in Australia, and even now, when writing to me, he signs himself "a renegade plant pathologist."

In Queensland, the Entomologist [Henry Tryon] is also Vegetable Pathologist; and I understand that it is the same in Tasmania, but in these days of specialisation, from the nature of fungus diseases, it is quite impossible for any single individual to do justice to both divisions of the subject.

In 1906 I moved into my present quarters with six rooms. There is my own laboratory and one for my assistant, museum and library, herbarium and photographic room, dark room and storeroom. The work grew apace, and as numerous field experiments were being conducted, only the more important diseases could receive close attention. In connexion with the work of the Wheat Improvement Committee, one eighth of an acre at the Burnley Gardens was enclosed with bird-proof netting, so that experimental plots could be harvested without loss.

6.5 Some of the Work Accomplished

It is impossible in a short article like the present, to give an adequate idea of the work done during the past 20 years. It is partially recorded in Annual Reports and in the various works and pamphlets published during that period, containing the results of field experiments as well as of laboratory investigations.

The black spot of apple and pear received early attention, for as one of our largest orchardists remarked to me, "I will either have to conquer the Black Spot or Black Spot will conquer me." In these early days there was also the prevalent belief to contend with that the disease was entirely due to the weather, and when I was experimenting in the treatment of the disease another orchardist bluntly told me that he had no faith in securing a remedy, for "when you can control the weather, you can control the Black Spot." Fortunately this prejudice has been largely overcome, although history is repeating itself in the case of Irish Blight, and I am plainly told by some potato growers that the disease is in the air, that when you

Fig. 6.1 McAlpine, Departmental Officers and Farmers inspecting a Field of Potatoes (Permission, State Library of Victoria, Melbourne – La Trobe Australian Manuscripts Collection [MS 12772, Box 3555/3])

have a heavy rainfall at the growing season followed by sweltering heat, what can you expect! [McAlpine overcame the ignorance about the origin of diseases such as Irish Blight through lectures and field demonstration where he enlisted support from Departmental officers (Fig. 6.1) DGP].

However, following the lines of American investigators, Black Spot has now been overcome by spraying with Bordeaux mixture at the right season, and this is so important, just to catch the fungus when the spores are germinating, that a single spraying, thoroughly done, with a properly prepared mixture or solution, is generally sufficient."

The Stinking Smut of wheat no longer troubles the careful farmer, since he knows that by properly steeping his seed, his crop is clean, and quite recently the treatment of maize seed has been found effectual in preventing smut. Some puzzling diseases, such as "Take-all" in wheat, and "Fairy rings" in bowling greens and lawns have been traced to their causes.

The diseases of the principal products grown have been specially studied and the results published, such as those of stone fruits and citrus trees, vines, potatoes, cabbages and cauliflowers, etc. The latest work just issued is *The Smuts of Australia* in which various smuts of cereals are thoroughly dealt with as well as those of native grasses.

During the past 20 years diseases of every description of plant have been sent to me from all parts of Australia, and experiments have been carried out with various modes of treatment. The results will be given in a work, which will appeal to the man on the land, under the title of *Australian Plant Diseases*.

[An explanation of why this work never appeared is given at the conclusion to this article, DGP].

6.6 Problems Awaiting Solution

The Rust in Wheat problem is still with us, but experiments are under way for determining how far different species and sub-species of *Triticum* are rust-resistant. Then, by crossing a rust-resistant strain of inferior quality with some of our Australian wheats, it is hoped to secure a marketable wheat [variety] that will resist the rust, even when weather conditions are favourable to it. [He also had to distinguish between the different rusts that occurred on individual cereals (Fig. 6.2) DGP].

The Bitter Pit of the apple is another disease, which has baffled all our efforts at prevention, but since it is more or less common to all States, there is every likelihood of a joint action being taken in the scientific investigation of its cause and

Fig. 6.2 Wheat leaf and stem rusts by Brittlebank. From McAlpine's *Rusts of Australia* (1906), Plate A (Permission of the Department of Environment and Primary Production, Victoria)

treatment. Just as the medical men, after all these years of investigation and research, have as yet failed to deal effectively with cancer and consumption, so the plant pathologist has still such diseases as bitter pit of the apple and black rust of wheat to overcome, but the very fact that scientific method is being brought to bear on their study is a hopeful sign of their ultimate control. The employment of parasitic fungi for the destruction of insect pests is a well worthy of immediate attention and we already know fungi which prey upon scale of the orange and the larvae of the codlin moth in Australia.

The diseases of our native timber trees are still practically untouched and the knowledge of bacterial diseases of plants is still in its infancy. The parasitism of bacteria has been proved equally with that of fungi and it will be necessary in the immediate future to have a bacteriologist attached to the branch.

6.7 How This Study Has Benefited the Producer

Whatever makes for health makes for quality, and it is a self-evident proposition that success depends on avoiding losses as well as on securing profits. Poor crops and stunted growth are evidence either of unsuitable conditions, or of want of knowledge on the part of the grower. It is the business of the plant pathologist to help the producer to gain the full reward of his industry and skill by guarding against the unseen foes, which too often rob him of the fruits of his labours. It is not always possible to put a monetary value on the gains secured by preventive treatment, but when a farmer sees his promising wheat crop ruined by rust or the orchardist his fruit crop considerably reduced by the ravages of black spot, he has no difficulty in summing up his losses. Still, reliable statistics are not available in Australia to show the actual losses over a wide area, but in other countries attempts have been made to reckon it up, and the results are sufficiently startling. In Germany the losses on the potato crop for 1908, through planting the diseased tubers of 1907, was reckoned to be at least £30,000,000. In the State of California it was estimated that in the single disease of peach leaf curl alone, £80,000 was saved in 1 year, by spraying at the proper time. In the United States it was calculated that the annual loss by Oat smut was over £3,300,000 and in New York District, Irish Blight was held responsible in 1904 for a loss of over £2,000,000.

[*NB According to the Reserve Bank of Australia, the value of £1 sterling, the Australian currency of 1910, in 2012s currency is equivalent to $ Aus.144.55*]

When it is considered that this loss is largely preventable, and that field demonstrations have shown, in many instances, how it may be avoided, it will be seen that not only the individual farmer but also the State as a whole, would be the gainers, if the precepts of plant pathology were put into practice.

If it be asked why the producer, who is generally blind to his own interests, does not adopt the methods which have been proved to produce such good results, then I can only reply in the words of Samuel Johnson, who when asked by a lady why he defined *pastern* as *the knee of a horse*, replied "Ignorance, madam, pure ignorance," or perhaps it is ignorance and apathy combined.

6.8 Conclusion

Plant Pathology is now a recognised science, just as much as human or animal pathology, although it is only since 1866 when De Bary published his work on the Morphology and Physiology of Fungi, that the study has been advanced sufficiently and placed on such a sure foundation, as to come within the circle of the sciences. I still look forward to the time when a central laboratory will be established by the Commonwealth for the study of plant diseases, where the most complete equipment will be provided for investigating the nature and life history of the organisms concerned in causing disease, where the best men obtainable will be engaged in the study of the physiology as well as the pathology of the plant-life where investigations will be carried out in a systematic manner dealing with the various problems which arise, especially such as require special skill and prolonged research for their elucidation.

While the laboratory work would thus receive its due share of attention and concern itself with those diseases, which are common to all States. Or such as no individual State could profitably undertake with the resources at its command, it would be supplemented by field experiments conducted in the various States and with their co-operation. These experiments would be designed more particularly for testing methods of prevention or at least of mitigating the injurious effects of the parasite. Such an institution would leave the States free to deal with their own local diseases and it could be made a teaching adjunct to any existing schools of agriculture. Plant pathology would thus become a recognised subject of instruction for Agricultural students, just as human pathology is for medical and animal pathology for the veterinary student.

At the present time there is no provision made for such a course and this would form the coping-stone to a system of agricultural education, in which the plant itself would be the main object of study, both in the normal state of health and the abnormal condition of disease. Besides, there are no facilities offered for the training for those who have to administer the Quarantine Act and detect disease in imported plants or parts of plants. The recent importation of French lucerne seed, badly infested with spores of the rust, shows the urgent necessity for something being done in this direction. Further, when some of the States wished to appoint plant pathologists, more particularly with reference to potato diseases, there were none available, and this is another strong reason for providing the means whereby men may be properly trained for the work, in all the methods of pathological research, both in the laboratory and the field. Farmers and fruit-growers are beginning to realise that disease may be controlled, or at least mitigated, and that it is necessary to produce the best possible crop free from disease at the least possible cost to meet increasing competition. Even in the United States, with its numerous agricultural colleges, where plant pathology is part of the course, Dr. Erwin F. Smith remarks that at present and for some time to come the demand for well-trained plant pathologists is likely to be well in excess of the supply.

Now that irrigation is about to be extensively introduced and closer settlement is the policy of the States, it is more than ever necessary to attend to the diseases caused by fungi, for moist conditions combined with heat are very favourable to their development and the contiguous planting of similar crops favour their spread and if unchecked may result in epidemics."

6.9 International Recognition

By 1910, McAlpine's reputation was solid in Britain, Europe, and North America. Appeals for his assistance were made from Africa, Japan and elsewhere while he regularly exchanged news and ideas with a network of correspondents around the world. His reputation was founded among mycologists and plant pathologists in Australia and internationally, and Colleges of Pharmacy and Schools of Veterinary Science here and abroad. By 1917, his recommendations for reducing the incidence of bitter pit were greatly appreciated and highly praised in Europe and the United States by staff in Colleges, Universities and State Departments of Agriculture as well as those engaged in horticultural industries.

In 1891, *Zeitschrift für Planzenkrankheiten* the first European journal devoted to plant pathology, began publication. On its title page (Fig. 6.3), where its purpose is defined, it is described as an organ for those interested in plant protection. It was produced with the assistance of an *International Phytopathological Committee*, consisting of over 44 eminent world scientists, representing their countries worldwide. Daniel McAlpine name headed the alphabetical list containing such historic figures as Kühn, Millardet, Penzig, Plowright, Prillieux, Saccardo, Marshall Ward and Woronin. Professor H.H. Whetzel (1918) claimed McAlpine as one of the most important plant pathologists of the Millardetian period, 1885–1906. The prestigious German scientific society, the Caesarean Leopold-Caroline Academy of Natural Phenomena, made him an honorary member in 1894; and in 1919 he was admitted to Foreign Membership of the Royal Swedish Academy of Agriculture on the same day as Professor Jakob Eriksson, an expert on rust diseases (White 1986).

Was it any wonder therefore that McAlpine was reluctant to step outside his familiar, comfortable and greatly enjoyed domain in plant pathology into one which was unfamiliar and mostly unexplored and unknown. It may have been the thought of leaving the familiar and being drawn into the unfamiliar that resulted in an address to the Field Naturalists entitled *The Romance of Plant Pathology* (McAlpine 1910c).

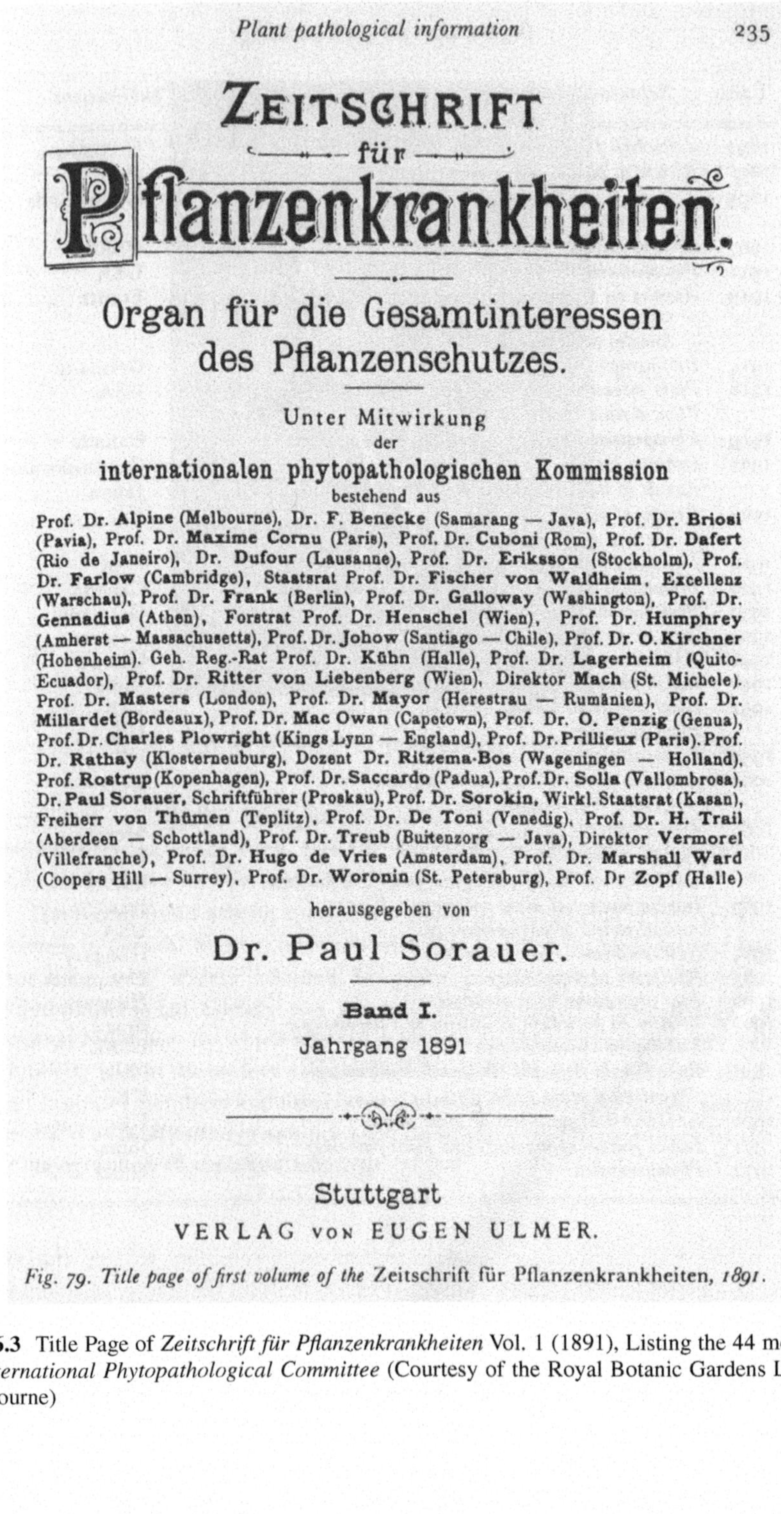

Plant pathological information 235

ZEITSCHRIFT

für

Pflanzenkrankheiten.

Organ für die Gesamtinteressen
des Pflanzenschutzes.

Unter Mitwirkung
der
internationalen phytopathologischen Kommission
bestehend aus

Prof. Dr. **Alpine** (Melbourne), Dr. **F. Benecke** (Samarang — Java), Prof. Dr. **Briosi** (Pavia), Prof. Dr. **Maxime Cornu** (Paris), Prof. Dr. **Cuboni** (Rom), Prof. Dr. **Dafert** (Rio de Janeiro), Dr. **Dufour** (Lausanne), Prof. Dr. **Eriksson** (Stockholm), Prof. Dr. **Farlow** (Cambridge), Staatsrat Prof. Dr. **Fischer von Waldheim**, Excellenz (Warschau), Prof. Dr. **Frank** (Berlin), Prof. Dr. **Galloway** (Washington), Prof. Dr. **Gennadius** (Athen), Forstrat Prof. Dr. **Henschel** (Wien), Prof. Dr. **Humphrey** (Amherst — Massachusetts), Prof. Dr. **Johow** (Santiago — Chile), Prof. Dr. **O. Kirchner** (Hohenheim). Geh. Reg.-Rat Prof. Dr. **Kühn** (Halle), Prof. Dr. **Lagerheim** (Quito-Ecuador), Prof. Dr. **Ritter von Liebenberg** (Wien), Direktor **Mach** (St. Michele). Prof. Dr. **Masters** (London), Prof. Dr. **Mayor** (Herestrau — Rumänien), Prof. Dr. **Millardet** (Bordeaux), Prof. Dr. **Mac Owan** (Capetown), Prof. Dr. **O. Penzig** (Genua), Prof. Dr. **Charles Plowright** (Kings Lynn — England), Prof. Dr. **Prillieux** (Paris). Prof. Dr. **Rathay** (Klosterneuburg), Dozent Dr. **Ritzema-Bos** (Wageningen — Holland), Prof. **Rostrup** (Kopenhagen), Prof. Dr. **Saccardo** (Padua), Prof. Dr. **Solla** (Vallombrosa), Dr. **Paul Sorauer**, Schriftführer (Proskau), Prof. Dr. **Sorokin**, Wirkl. Staatsrat (Kasan), Freiherr **von Thümen** (Teplitz). Prof. Dr. **De Toni** (Venedig). Prof. Dr. **H. Trail** (Aberdeen — Schottland), Prof. Dr. **Treub** (Buitenzorg — Java), Direktor **Vermorel** (Villefranche), Prof. Dr. **Hugo de Vries** (Amsterdam), Prof. Dr. **Marshall Ward** (Coopers Hill — Surrey), Prof. Dr. **Woronin** (St. Petersburg), Prof. Dr **Zopf** (Halle)

herausgegeben von

Dr. Paul Sorauer.

Band I.
Jahrgang 1891

Stuttgart
VERLAG VON EUGEN ULMER.

Fig. 79. Title page of first volume of the Zeitschrift für Pflanzenkrankheiten, *1891*.

Fig. 6.3 Title Page of *Zeitschrift für Pflanzenkrankheiten* Vol. 1 (1891), Listing the 44 members of *International Phytopathological Committee* (Courtesy of the Royal Botanic Gardens Library, Melbourne)

References

McAlpine D (1910a) Twenty years of plant pathology in Australia. Report of the Vegetable Pathologist 1907–1910. Department of Agriculture Victoria, Government Printer, Melbourne, pp 56–61

McAlpine D (1910b) Report of the vegetable pathologist 1907–10. Department of Agriculture Victoria, Government Printer, Melbourne, pp 44–61

McAlpine D (1910c) The romance of plant pathology. Victorian Nat 27:127–129

Whetzel HH (1918) An outline of the history of phytopathology, with 22 portraits. W.B. Saunders Company, Philadelphia/London, 130 pp

White NH (1986) McAlpine, Daniel (1849–1932). In: Australian dictionary of biography, vol 10. Melbourne University Press, Melbourne, pp 193–194

Part II
The Bitter Pit Investigation

In his 20 years as Victorian Vegetable Pathologist, McAlpine had published 218 papers, bulletins and books and had three further papers to be published in 1912, in preparation or in press. By 1921 he had published 230 articles of various kinds. He published no more on mycology after 1912. Sadly, 1911 spelled the end of the career that he loved as a mycologist/plant pathologist. The dearth of trained plant pathologists, referred to in his report, resulted in his being drawn into a 5-year search for the cause and means of control of the disorder of apples, pears and quinces known as bitter pit. In a sense, McAlpine designed the trap that drew him into the investigation. His concern about bitter pit was noted early in reports and papers prepared or published prior to the investigation. He described and compared the vascular systems of apple pear and quince fruit in an attempt to gain some understanding of the cause of bitter pit because he believed the vascular system was somehow involved. He summarised the history of bitter pit, and set out his views on the need for Commonwealth and State co-operation in resolving Australia-wide disease problems. While his views were expansive and far-sighted, in a sense McAlpine was hung on his own petard. By 1917, his career, which from 1876 to 1910 was no less than illustrious, was over and he was treated in a manner he most certainly did not deserve.

At the height of his fame and influence, McAlpine was effectively removed from the occupation that had absorbed him for 20 years. Against his wish, but believing he was only being seconded to the Commonwealth for 4 years, he reluctantly accepted appointment to investigate the cause and control of a highly destructive disease of apples known as 'bitter pit'. McAlpine recognised that bitter pit was a physiological disorder, probably resulting from a number of interacting factors, some within the plant, some within the soil and all moderated by seasonal conditions. This way of thinking was ahead of its time. He changed from his usual approach to resolving disease problems. During the previous 20 years he had strongly advocated a two-pronged approach to diagnosing plant diseases and devising control measures for them. First, he believed it essential to know as much as possible about the healthy host-plant before assessing its responses to a disease and second, to identify the pathogen and understand its biology in order to decide on an appropriate control measure. In the case of bitter pit, however, because he believed it was a physiological disorder, he evidently decided that the best use of his limited time was to evaluate the effectiveness of orchard management practices considered by the best orchardists and experienced orchard inspectors to be effective in reducing the problem. He did this very effectively through improvements to pruning regimes, fertiliser programs, moisture control, the identification of the least susceptible apple varieties, and the immediate cool storage of harvested fruit. Through his recommendations, he substantially reduced the incidence of bitter pit and vastly improved the quality of fruit exported to overseas markets. He won the universal gratitude of orchardists and traders and high praise from overseas, but not universally in Australia.

Chapter 7
Bitter Pit

7.1 History of Research into Bitter Pit

Bitter pit is evident on affected fruit, particularly apples, as small sunken pits in the skin under which the flesh is dead, brown and bitter (Figs. 7.1 and 7.2). These are most commonly noticed first at the calyx (flower) end of the fruit and progress toward the peduncle (stalk). As more cells die the flesh of affected fruit gradually turns brown, progressing from beneath the skin towards the core. Sometimes fruit becomes irregular as the disease distorts it. Cobb (1903) referred to this as 'stigmonosis', a term later also used by McAlpine (Fig. 7.3). As the extent of dead cells increases the fruit becomes inedible and of no use in any other way. For reasons still not fully understood, the incidence of the disorder increased dramatically over the last decades of the nineteenth century. In part this may have been due to the gradual depletion of certain soil minerals and increases in soil acidity. The condition often severely limited the commercial yield of trees of susceptible cultivars, especially if lightly laden with oversized fruit. The incidence of bitter pit was increased by picking fruit before it was fully mature, and by not storing it under cool conditions. When apparently sound fruit was picked and sent to market, there was still little guarantee that it would be edible after a month in local storage or shipment to Europe. Fruit with bitter pit was unsaleable. In 1910, McAlpine (1910a) reported a trial at Burnley in which, at harvest, 67.5 % of fruit was pitted and after a further 26 days in storage, 90.7 % was pitted and unsaleable. This was not uncommon in localities where the incidence of bitter pit was high and the most susceptible varieties were grown. It was disastrous for orchard businesses reliant on varieties such as *Cleopatra* or *Annie Elizabeth* in localities conducive to bitter pit.

Just when bitter pit first appeared is unknown. It possibly appeared or was noticed following the domestication of the apple and the development of orchard husbandry, but it was not recorded before the closing decades of the nineteenth century. Bünemann (1972) claimed that Jaeger (1869) published the first report of bitter pit to appear in the scientific literature, concluding that the spots were caused

D.G. Parbery, *Daniel McAlpine and The Bitter Pit*,
DOI 10.1007/978-3-319-09552-3_7

Fig. 7.1 Painting of Bitter pit by Margaret Senior. From Agfact H4.AC.1, Second ed. J.A.Dart 2004 (With permission of the NSW Department of Primary Industries, Images collection. Courtesy Lowan Turton)

by some kind of fermentation. However, Jaeger was not referring to bitter pit of apples but to 'Stippigwerden der Kernobstfrucht', which translates as 'spotting of stone fruit', possibly apricot freckle! The German name for bitter pit was most commonly 'Stippen' or 'Stippenflecke'. In a lengthy letter written in 1909, McAlpine (1909b) told the Hon. George Graham that another German, Sorauer, had first described bitter pit in 1879 from apples found near Berlin, and that the German name of *Stippen* came from the 'stippling' used by artists. From Cobb's (1895) observations given in his report to the Pomological Society, however, bitter pit appears to have been noted in North America earlier than 1875. A former government inspector of orchards in New Zealand stated that it had been noted in New Zealand since the early 1860s (McKenzie 1911). It is most likely, however, that bitter pit had been around for many years without being distinguished from other diseases. In a letter written by Thomas McKenzie (1911), the Engineer at the Cape Butter and Bacon Factory in Wynyard, Tasmania to The Hon. George Graham,

Fig. 7.2 Bitter pitted Apples. Agfact H4.AC.1, Second ed. J.A.Dart 2004 (Permission of the NSW Department of Primary Industries Images collection, Courtesy Lowan Turton)

McKenzie referred to the presence of bitter pit in an orchard associated with Dalkeith Castle in Scotland in the 1840s.

Prior to McAlpine's involvement in the problem, some fundamental observations had been made. Wortmann (1892) concluded that the disease occurred under conditions of abnormal transpiration. He noted that apple trees in which water was conducted readily from lower cells to replace that lost through transpiration were least susceptible to *stippen* (bitter pit). Cobb (1895) made a number of important observations when he reported the unusual retention of starch in affected cells within the pits, in addition to noting the influence of climate and varietal differences on incidence.

During 1910, John Osborne (1911) who was the Fruit Expert for the Tasmanian Government made an extensive and valuable review of what was known about bitter pit for the Fruit Grower's Association of Australasia. His conclusions agreed with McAlpine's views at the time. He favoured the explanation of the cause given by Pole Evans (1909) who made a detailed examination of the anatomy of the pits and drew the following conclusions:

> These facts lead me to conclude that this unhealthy condition of affairs is brought about by the following sequence of events. The cells surrounding the vascular bundles, or those commonly situated at their ends, are being continually subjected to a great internal pressure, with

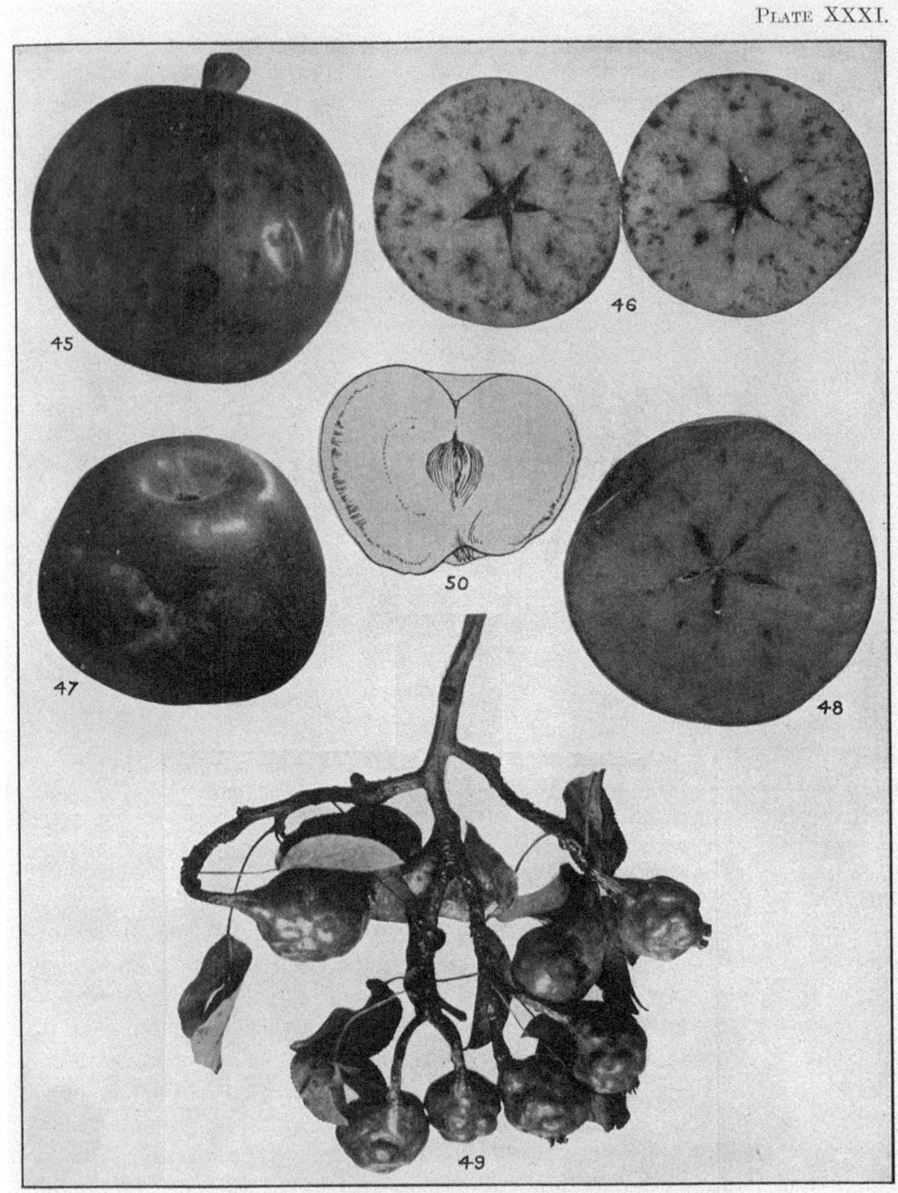

Fig. 7.3 Stigmonosis in Apples. McAlpin's Second Progress Report on Bitter Pit, Plate XXXI (1913) (Courtesy of the Royal Botanic Gardens Library, Melbourne)

the result that they become enlarged and thicker walled. This internal pressure is undoubtedly due to an accumulation of water, which inflates the cells to such a pitch that in many cases the tension becomes so great that the cells can no longer stand the strain put upon them, and consequently burst. As soon as the cell is ruptured atmospheric oxygen gains access to the

> cell, and together with the enzymes present act on the tannin, producing dark-coloured oxy-compounds, which are precipitated on the walls as a gummy substance while any further diastatic action is inhibited, with the result that the starch grains remain unaltered, and are always abundant in the vicinity of the necrotic areas. As soon as a few cells bordering on the vascular bundles, or those situated at their ends, are ruptured, their neighbours surrounding them and depending on them for their water supply are immediately affected, and consequently suffer also with the result that a nest of unhealthy and dried-out cells becomes one of the dry and tough bitter spots. The collapse of these cells in the interior of the fruit eventually leads to the characteristic depression or pit on the exterior.

According to Osborne (1911), Evans (1909) explained that water pressure increased as a result of cool night temperatures reducing the rate of transpiration, while soil temperature remained high enough to maintained root function, suggesting roots absorbed and translocated water faster than leaves and fruit transpired it. In the absence of knowledge about plant physiology at the end of the twentieth century, there was evidently enough plausibility in Evan's hypothesis for many scientists to accept it.

The extent of concern over the problem might be gauged from the numbers of papers published about it. Over the three final decades of the nineteenth century, 12 papers were published. During the first decade of the twentieth century there were 21 papers, the second saw 50 papers; during 1931–1940 there were 57 papers; 1941–1950 – 81 papers; 1951–1960 – 78 papers; and 1961–1970 – 400 papers plus another 45 to the end of 1972 (Bünemann 1972). Thus over the century from 1879, in the order of 800 papers about bitter pit were published. Even in 2011, 100 years after the setting up of a commonwealth research effort led by Daniel McAlpine, while many of the factors involved in causing bitter pit are understood and the disorder is generally well controlled, horticulturalists are not fully confident that there is absolute understanding of its cause. It was not until 1978 that the present state of knowledge about the cause was within reach, and the most effective means of avoiding it determined.

7.2 Bitter Pit in Australia

In 1886 Crawford referred to 'spotted' apples in a Government Report on diseases of apples in South Australia. This is believed to be the earliest Australian record of bitter pit (Crawford 1886), although W.C. Grasby (1911) reported seeing it on *Cleopatra* apples in his father's South Australian orchard in about 1880. L.R. Jones first recorded it in Victoria in 1891 on *Baldwin* apples grown at Vermont (Carne et al. 1929), while Cobb (1895) recorded it in NSW in the early 1890s and coined the name *Bitter Pit*.

Bitter pit is a physiological disorder of apples, pears and quinces. Nathaniel Cobb (1895) coined its English name because of the bitter flavour of the dead flesh beneath the pits that develop in the surface of affected fruit. Bitter pit caused severe economic stress in the apple and pear industries in Australia and elsewhere from the late nineteenth century until well into the twentieth century. While it is still found, it no longer causes the disruption and economic hardship that it did. The fact that it is no longer the threat that it was over 120 years ago is due to the work of hundreds of scientists worldwide, who have gradually unravelled much of the complex physiology, identified many of the interacting factors involved, and have identified environmental conditions conducive to its onset. McAlpine understood the importance of orchard management and storage requirements. He used his knowledge to good effect to substantially reduce bitter pit's impact. Today's control protocol still incorporates all his recommendations. It was not until 1954, however, that researchers identified the roles played by key nutrients in the physiology of pome fruits in causing the disorder. Physiologists, however, are still not confident that they fully understand the combinations of genetic predisposition to bitter pit, with the influence of nutrients availability and balances on the physiology of developing fruit, and the overriding effects of water availability, temperature, and crop density (Little and Holmes 2000; Dr. R. Holmes, 2010, personal communication).

The problem had become so great by the first decade of the twentieth century that it was brought to the attention of every State Premier and ultimately the Prime Minister. Public awareness was heightened by the proposal that bitter pit resulted from the use of an insecticide to control the introduced pest, codlin moth. Awareness turned to public concern when newspapers such as *The Age* and *The Argus* in Melbourne reported the linking of codlin moth control to the emotive word 'poison'. Bitter pit then became a matter of public interest, concern and considerable debate. Most orchardists rejected the poisoning proposition on the grounds that bitter pit was badly affecting some apple varieties many years before the use of spray to control codlin moth. At the same time, the general public was concerned that if sprays were affecting the fruit, they may also be affecting those who ate it. Bitter pit was affecting the livelihoods of orchardists, stock and station agents, shipping agents and exporters, and an unfounded scare over a potential poisoning of apple-eaters was likely to seriously compound their problems.

7.3 The Economic Importance of Apples

In 1891 the first export of Victorian apples realised £815 for c.3,250 bushel- cases of fruit. At first, fruit was packed in standard sized wooden kerosene cases. These were made to hold two four-gallon square-based tins of kerosene, i.e. 8 gallons or 36,363 cc. In 1909, the use of kerosene cases for packing fruit was banned and the volume of the bushel case was standardised to 2150 cubic inches [35,232 cc], a reduction of 3 % in volume from the kerosene case. The bushel being a volumetric measure, however, the weight was slightly variable according to the size and water content of the fruit. There was, however, no international standard weight for a bushel of apples. Britain and New Zealand considered a bushel of apples was 40 lbs [18.2 kg], Australia 42 lbs [19 kg], Canada 45 lbs [20.5 kg] and the USA 48 lbs [21.8 kg]. In the following, where bushels are expressed in weight, 42 lbs per bushel is used. As a guide to the number of apples in a bushel case, McAlpine's rule of thumb was two large apples, three medium sized apples and four small ones equaled one pound's weight. When Australian orchardists noted that 314 tons [324 tonnes] of apples could fetch £815, that is £2/11/10 per ton, they saw great export opportunities for the industry and an accelerated expansion of orchards began, which continued for many years.

In spite of seasonal differences, but with increasing acreages, the annual increase in production from 1891 to 1900 was 26 % and from 1901 to 1910 it was a further 53.6 %. By the time agitation for government action over bitter pit began in earnest, commercial apple production in Victoria had reached 1,121,702 bushel-cases [21,033 tons] worth £560,351, realising £26/12/8 per ton, a 123 % increase in apple prices over the two decades. Over the eight relatively bitter pit-free years between 1904 and 1915 the Victorian average annual apple production was 1,285,219 cases [24,098 tons] worth £642,605, whereas for the 3 years of 1904, '05, and '06, when there was very high incidences of bitter pit, the average yield of saleable fruit was 568,940 cases [10,668 tons] worth only £284,470, 55.7 % down on the years when there was a low incidence of bitter pit. Between 1904 and 1915, the price of apples was maintained at £26/13/4 per ton. The retail price today is equivalent to £1,150 per ton.

By 1910, Tasmania, Victoria and New South Wales produced close to 70 % of Australia's apple exports of approximately 3,856,000 cases [72,305 tons] worth £1,928,000 and South Australia, Western Australia and Queensland produced ca 30 %, i.e. 1,652,570 cases [30,988 tons] worth £826,286. Thus the total Australian export trade was in the region of 5,508,570 cases [103,293 tons] worth £2,754,285. Today, just 100 years later this equates to approximately $800 million. Apples had become an important contributor to the national economy. This demonstrates the reason for concern about bitter pit, which in bad years (Carmody 1910; Turner 1910) reached such proportions as to reduced harvests by over half. Such events in the first decade of the twentieth century culminated in State Ministers of Agriculture calling on the Commonwealth Government for assistance.

7.4 An Increasing Problem 1886–1910

In 1890, with the aftermath of the 1889 wheat rust epidemic ringing in the treasury's ears, the Victorian Government appointed Daniel McAlpine as its Vegetable Pathologist and, shortly after, the New South Wales Government appointed Nathaniel Cobb. They were equally distinguished. Cobb was from the USA where the first vegetable pathologist was appointed in 1885. Cobb also made important contributions to plant pathology in Australia and later in the USA after he returned there in 1905. During the period from 1890 to 1905, McAlpine and Cobb were called on for help with plant disease problems Australia-wide and even further afield. During 1898–1901, while Cobb was overseas, and again from 1905 until 1911, Daniel McAlpine was the only plant pathologist in Australia south of Brisbane. He competently handled an enormous workload, which included inquiries from Western Australia, South Australia and Tasmania, where such appointments were not made until 1911, 1912 and 1913 respectively.

Almost from the time of their appointments, Cobb and McAlpine were made aware of the deepening concerns of orchardists over bitter pit and its ruinous financial consequences. During the decades prior to their appointments, apple and pear orchards were seriously afflicted by infestations of harlequin bugs and codlin moth. With the introduction of lead arsenate sprays in 1892, however, these were brought under control, possibly giving greater exposure to the higher than expected incidence of bitter pit, of which many orchardists were aware, especially in the south eastern states. In about 1893, the Pomological Society of NSW requested Cobb (1895) to investigate the nature and cause of the disease and advise fruit growers on the best means of treating it. Cobb stated that he had been aware of the disease for at least 20 years, noting that it did least damage in cool climates, was worst in tart varieties and in late maturing rather than in early varieties. He commented that good keeping varieties were seldom affected. Cobb expressed the view that its cause was more to do with the tree than the fruit itself. Interestingly, while he ruled out filamentous fungi and initially insects as causes, he did not rule out non-filamentous microbes and, according to Smith (1911), Cobb later considered the bites of insects such as mosquitoes could possibly be involved. Viruses were not considered, probably because it was only five or so years since the discovery of Tobacco Mosaic Virus and, if thought about at all, the possibility was considered unlikely.

In the years immediately following their appointments, Cobb and McAlpine were not only confronted by the enormous problems of cereal black stem rust and bunts and smuts, which took much of their attention, but also spent greater amounts of time addressing increasing concerns about bitter pit, especially in apples. Between 1895 and 1903, Cobb published four papers on the topic (Cobb 1895, 1897, 1898, 1903) and in McAlpine (1904b) reported on experiments with manurial treatments that he had conducted at Pakenham with the susceptible cultivar *Sturmer*, since 1902. No satisfactory results had been obtained and he was continuing the trial as well as extending it to other districts. The problem was taking up increasing amounts of their time and resources.

7.5 A National Investigation Proposed

On 5th January 1909, McAlpine (1909c) sent a letter to Mr. Duffus, Secretary to the South Australian Department of Agriculture, pointing out that there was no concerted action between Officers and Ministers of Agriculture of the interested States in mounting a national investigation into bitter pit. The States he named were Victoria, New South Wales, South Australia, Western Australia and Tasmania. He also pointed out that he did not have time to engage in the investigation himself without putting aside urgent experiments on potatoes, and his commitment to the Wheat Improvement Committee. He considered that an investigation of the cause of bitter pit required a full time officer, and expressed the view that the scale of the investigation required the involvement of the Commonwealth Government.

Ironically, the letter to Duffus sowed the seeds of his own demise. McAlpine also relayed the conclusions of Professor J.B. Farmer D.Sc. FRS of Imperial College London, that bitter pit was a physiological disorder and the involvement of fungi or bacteria was considered unlikely. This had confirmed earlier advice received in 1903 from Dr Bull, bacteriologist at the University of Melbourne, to whom he had provided samples of pitted apples (McAlpine 1904a). Later, on 17th January, McAlpine (1909d) wrote to the Victorian Minister for Agriculture, George Graham, again expressing his view that the investigation of bitter pit was the duty of a National Department of Agriculture. He recognised the difficulties inherent in setting up such an investigation, but urged it be implemented.

Six months later, various fruit grower associations began pressing Minister Graham to take action to resolve the bitter pit crisis. For example, on 2nd June 1909 the Geelong and Western Districts Fruit Growers' Association wrote advising him of their resolution to urge the Government to appoint an expert to inquire into the cause of bitter pit and to find a remedy for it (Anon 1909). Ten days later this was supported in a letter written by Dennys, Lascelles, Austin & Co'y (1909) Wool Brokers, Stock, Station and Shipping Agents, Geelong. Their letter sent to Minister Graham, on behalf of the Geelong and Western District Fruit Growers' Association, supported the growers' resolution. Adding to this, the Secretary of the Chamber of Agriculture conveyed a resolution to the Secretary of Agriculture and Treasury from a meeting of Fruit Growers and Traders held at Bendigo, stating: 'that the Chamber impresses on the Government the seriousness of bitter pit in fruit, and urges the Department of Agriculture to use its best endeavours to investigate the disease with a view to its suppression.'

In the meantime, the Victorian Minister of Agriculture had asked McAlpine for a report, as detailed as possible, on what was known about bitter pit in Australia. McAlpine (1909d) sent him a six-page letter referring to the bitter flavour of most pitted apples and especially of pears, which he said tasted of quinine. He noted that Sorauer (1909) had first described bitter pit in 1879 from apples near Berlin, and that the German name was *Stippen*. He expressed his view that the investigation was the duty of a National Department of Agriculture, and although he recognised the difficulties he urged that it be got on with. He drew attention to the fact that in varieties such as *Bismarck*, pit developed in up to 90 % of fruit still on the tree, while in others, such as *Jonathan*, fruit appeared sound when shipped but pit developed during voyages to overseas markets. He noted its occurrence in Germany, Canada, USA, and the Cape of Good Hope in addition to the worst affected Australian states of NSW, Victoria, Tasmania and South Australia. He told the Minister that scientists were still in the dark in understanding the cause, which hindered attempts to find a control. He added however, that he believed it was a physiological condition and, as such, possibly involved a variety of causes. He ended by saying that the resolution of the problem required the energies of a trained specialist and increasing losses each year signaled the need for an exhaustive investigation without delay. The existence of this hand written report became known and copies were requested. The Australian Produce Company received their copy within 5 days (Plaff 1909).

7.6 The Perth Conference

By the end of 1909 support for action at the national level was strengthening. The National Fruit Growers Conference of Australasia was held in Perth from 26th to 28th October 1909, chaired by Mr L.M. Shoobridge[A1], President of the National Fruit Growers' Association. This might be seen as the official launch of a determined national effort to overcome the problem of bitter pit. It was a high profile affair opened by the Governor of Western Australia, Sir Gerald Strickland, the Premier, The Hon. N.J. Moore[A1], and several Ministers and politicians, some from interstate. Also present were delegates from every Australian state and New Zealand. In addition there were three official visitors, Mr J. Osborne Jnr., Horticultural Instructor from Tasmania, Mr G. Quinn[A1], Horticultural Instructor from South Australia, and Professor Lowrie, Director of Agriculture for Western Australia. The conference covered a wide range of concerns within the industry including orchard practices, various Acts of Parliament, uniformity of interstate inspections, shipping and trade. Six papers were presented and discussed. Dr J.H.L. Cumpston[A1] read a paper on the importance of fruit in diets, and Mr A.D. Cairns, from the W.A. Government's Refrigeration Works, presented a paper on *The Cool Storage of Fruit*. The conference also adopted motions which had bearings on the investigation of bitter pit. The first expressed the opinion of the conference that the proposed Federal Agricultural Bureau should confine its operations to scientific research only. Three others called for a coordinated approach, under the auspices of the Association, to find out as much as possible about bitter pit, and stated concerns over impurities in some sources of orchard sprays. Mr L.M. Shoobridge[A1] (1909) forwarded a ten-page report of the conference to every State Minister of Agriculture, with a covering letter including the three resolutions relating to bitter pit. These were:

1. That the prevalence of bitter pit be brought under the notice of Ministers for Agriculture with a view to the appointment of a Board by each state interested, to consist of two practical growers and one scientist to report observations.
2. Each Board to collect all observations and information on bitter pit from growers and report them at the next National Conference. [The Boards for each state were appointed and members named in the report].
3. Called on State Governments to bring in [to law], a Pure Insecticide and Fungicide Act similar to the Victorian Government's Manures Bill.

Mr John Osborne Jnr undertook to investigate the problem and to report his findings to the following Conference scheduled for October 1910 in Hobart, which he did (Osborne 1911).

7.7 The Ministers Respond

In January 1910, Minister Graham wrote to the other states suggesting that a national scientific investigation into the cause of bitter pit should be set up, and that an application should be made to the senates of the nation's universities and to professors of biology and botany asking for advice and assistance in the matter. This was the first time since federation that such a nationally coordinated project had been contemplated. On 30th March 1910, Thomas Pascoe, Minister of Agriculture for South Australia, responded to Minister Graham (Pascoe 1910a) pointing out that Resolution 23 of the Conference of Ministers of Agriculture held in Melbourne in August 1909 affirmed the desirability of the states co-operating with a view to ascertaining the origin, nature and prevention of bitter pit. In a footnote added to Pascoe's letter on 5th April, Dr Cameron acknowledged that Victoria was ready to cooperate in regard to Resolution 23 of the August 1909 conference.

At the end of April 1910, Pascoe wrote to his Victorian counterpart, informing him that the southern states, Western Australia, South Australia, Victoria and Tasmania, had agreed to cooperate in the investigation and that he expected NSW would also agree (Pascoe 1910b). He posed four questions:

1. Should the cost be based on the areas in each state planted to apples and pears?
2. Should the Federal Government be asked to assist?
3. Is there a pathologist attached to any Department in your state or university who would be competent to undertake the study?
4. Should the investigation start forthwith?

Minister Graham then asked McAlpine for his view on Pascoe's letter in relation to questions 3 and 4, and on 5th May McAlpine (1910b) replied that regarding question three he knew of no one in Australia with the necessary training in Plant Pathology, but thought Professor Ewart might be consulted on the matter. Regarding question four, he replied 'the sooner the investigation started the better'.

Graham, acting on McAlpine's suggestion, asked for Ewart's advice. Professor Ewart (1910) responded in a manner that implied McAlpine was not the man for the job, but that he himself was! He wrote:

> For this investigation, the services of an ordinary technical expert are not likely to be of much service. An obscure disease of this kind will need a broader view and deeper insight for its solution than can be expected from the expert limited to one small branch of work'. He then puts in his own bid 'We therefore recommend a research Scholarship of say £300 *per annum* with travel expenses, with preference being given to a home or foreign university graduate with a thorough knowledge of plant physiology, plant pathology and chemistry.

The Government funded the scholarship, but ignored Ewart's advice regarding the appointment of an 'ordinary technical expert'.

In June, Graham (1910) wrote to Pascoe referring back to his letter of 30th April stating that he believed each state should make the same contribution and that the Federal Government might be asked to contribute towards the expense. He also explained that there was no pathologist in the Victorian Department or University who could undertake the work. He supported an early start on the investigation. Pascoe (1910c) responded that as there was no available vegetable pathologist in Australia, a request should be made to the Federal Government to obtain one from

Europe. This did not happen, and no record of its being attempted has come to light. Pressure was building on McAlpine, however, to accept the post. He was the first to propose the need of a national approach, he was known to the Directors of Agriculture, most Ministers of Agriculture, and widely respected among agricultural and horticultural officers in most states as well as traders in agricultural and horticultural products. He had an enviable reputation for investigating disease problems and in helping the man on the land combat them. He was highly respected, widely known and liked. Thus he tended to be the first person most people thought of as the one to solve the bitter pit problem. However, he was strongly focused on mycological problems and had no wish to change. Besides, on 13th September 1910, *The Argus* reported that while Minister Graham had been authorised to cooperate with other states in seeking an expert pathologist, he felt that Mr McAlpine could not be spared from his other official duties to study the disease throughout Australia. At the time, McAlpine did not wish to leave his post of Victorian Vegetable Pathologist and the important projects engaging his attention, such as the wheat rust project, potato diseases and his compilation of a handbook on Australian plant diseases.

In spite of what Minister Graham said on 13th September, he changed his mind. Later in September, McAlpine (1910c) responded to Graham's request to consider leading the investigation saying that in order to do so he would have to give up his private practice. It consisted mainly of his contract with the Victorian Government as Vegetable Pathologist, and his contract to provide lectures at the Victorian College of Pharmacy, both of which he thoroughly enjoyed and was reluctant to forego. In spite of his preference, since proposing it to the Ministers of Agriculture in South Australia and Victoria in January 1909, it is evident from his letter that he had given a deal of thought to how such an investigation might be managed. He outlined what he would expect to do, even detailing what his first report would contain. He noted that at the Perth Conference of 1909, New Zealand growers were also interested in such an investigation, and that a government official from the Cape stated that he believed the South African Union Government would be willing to contribute towards the cost of the investigation if McAlpine undertook it.

On 14th January 1911, J.P. Wilson, Minister of Agriculture for South Australia, wrote to George Graham (Wilson 1911) assuming that McAlpine would be available to conduct the investigation, providing sufficient inducement was offered. He believed a 5-year investigation should be allowed for. Wilson noted differences of opinion among State Ministers of Agriculture regarding whether or not to involve the Commonwealth. Some felt it would be best organised by the States alone, but considered it best to discuss it at the Conference of Ministers proposed for March. Wilson then proposed an alternative: (i) The Commonwealth Government be asked to contribute half the cost and the states of Victoria, New South Wales, Western Australia and South Australia contribute the other half; (ii) The States to contribute in proportion to the area under apples and pears; and (iii) the investigation be under the general supervision of a committee of representatives, one from each state and two appointed by the Commonwealth.

7.8 The Reluctant Appointee

Minister Wilson's three-point proposal was put to McAlpine who responded to Dr Cameron (McAlpine 1911) on 18th January 1911, as follows:

> I have considered the alternatives of retaining my connection with the State or serving the Commonwealth, and although the investigation of Bitter Pit under Commonwealth auspices is very tempting, yet on the whole I would prefer giving my services to the State, other things being equal. If I were appointed Vegetable Pathologist [as a public servant rather than consultant - DGP], the investigation of Bitter Pit would be part of my duties and the contribution of Victoria for that object would be saved and there would be no Commonwealth interference. Perhaps you are not aware that although New South Wales is stated in the South Australian correspondence as having agreed to join with the other States in this investigation, it is known to the Commonwealth officer who has charge of the matter that the Mother State will stand out. Then again if the Entomological Branch were re-arranged, the salary of the present Head would also be saved, and these two items would enable the Department, without incurring any extra expense, to retain my services. I am now 61 years of age and the age limit would not be a serious drawback.
>
> With reference to your suggestion of the advisability of having the different allied branches under one roof I may point out that there is plenty of room in the present building, presupposing that Dr Brown and his staff, whose duties have nothing in common with my own, were removed to more convenient quarters. With irrigation and Closer Settlement there will be plenty of scope for doing good work in my own particular Branch, and the preparation of the work on 'Australian Plant Diseases' treated in a manner suited to the man on the land, would be a fitting sequel to my 20 years experience in Plant Pathology in Victoria. Of course I understand that the Commonwealth offer of going halves with the States in their contributions would be satisfactory in a pecuniary sense, but my long connection with Victoria and a thorough knowledge of the conditions prevailing here would always be a strong attraction to continue the work already begun.
>
> I remain yours faithfully,
>
> D. McAlpine.

McAlpine wished to continue as the Vegetable Pathologist, but as a public servant, not a consultant. At various times, he referred to completing *Australian Plant Diseases*, a reference handbook for farmers, orchardists and gardeners, which sadly he never did. This letter seems to be an echo of a much earlier plea to a higher power! Is it McAlpine's 'let this cup pass from me?'

By February 1911, with pressure mounting on him from all quarters, it was widely expected as inevitable that McAlpine would lead the investigation, and even he was becoming resigned to his fate. Fifteen years later Plant Pathology, Entomology and Plant Physiology were merged into the Biology Branch in line with McAlpine's proposal.

References

Anon (1909) Letter [2-6-'09] from The Geelong and Western District Grower's Association to Minister Graham advising that the Association urge him to appoint an Expert to inquire into the cause of bitter pit and find a remedy

Anon (1911) Letter [23-5-'11] in *The Argus* newspaper suggesting the Government offer a reward of £1,000 to the person who discovers the cause or cure of bitter pit

Bünemann G (1972) Annotated bibliography on bitter pit of apples. Bibliographical series of the technological University of Berlin. No.2. University Library, Berlin, 170 pp, ISBN 3 7983 0106 9

Carmody PJ (1910) Report of the Chief Inspector of Orchards. Report of the Department of Agriculture, Victoria for the years 1907–1910. J. Kemp, Govt. Printer, Melbourne, pp 208–212

Carne WH (1911) Letter to Minister of Agriculture [15-5-'11] on behalf of the Gembrook South Fruit Grower's Association

Carne WM, Pittman HA, Elliot HG (1929) Studies concerning the so-called bitter pit of apples in Australia – with special reference to the variety "Cleopatra". CSIR Bulletin No. 41. Government Printer, Melbourne, 101 pp

Cobb NA (1895) Bitter pit of the apple. Agr Gaz NSW 6:859–861

Cobb NA (1897) Cause of an important disease of apple. Agr Gaz NSW 8:126–221

Cobb NA (1898) Brown or bitter pit of the apple. Agr Gaz NSW 9:683

Cobb NA (1903) Stigmonosis. Agr Gaz NSW 14:692

Crawford FS (1886) Report on Fusicladiums, Codlin Moth, and certain other fungus and insect pests attacking apple and pear trees in South Australia. Government Printer, Adelaide, 77 pp, cf p53

Dennys, Lascelles, Austin and Co. (1909) A letter [12-6-'09] supporting Western District Fruit Growers' call for the appointment of an expert investigator of the cause and control of bitter pit

Evans IBP (1909) Bitter pit of the apple. Transvaal Department of Agriculture, Technical Bulletin No. 1

Ewart AJ (1910) Letter [30-5-'10] to Department of Agriculture, Victoria, responding to inquiry as to who was qualified to lead the bitter pit investigation

Graham HG (1910) Letter [4-6-'10] to his South Australian counterpart responding to his letter of 30th April 1910

Grasby WC (1911) Letter [29-5-1911] written to *The Argus* Newspaper

Jäger G (1869) Über das Pelzig oder Stippigwerden der Kernobstfrucht. Illustrieren Monatschrift für Obstbau und Weinbau 318–319

Little CR, Holmes RJ (2000) Storage technology for apples and pears – a guide to production. Department of Natural Resources and Environment, Victoria, ISBN 0 7311 4466 X

McAlpine D (1904a) Correspondence relating to the examination of bitter-pitted fruit by Dr. Bull, Department of Microbiology, Melbourne University

McAlpine D (1904b) Report of the vegetable pathologist for 1903–1904. J Dep Agr Vic 12:1–9

McAlpine D (1909a) Letter [5-1-'09] to E.G. Duffus, Department of Agriculture South Australia, drawing attention to the lack of concerted action between officers of the different States, and Ministers of Agriculture of the interested States, to get on with a national investigation of bitter pit

McAlpine D (1909b) Letter [17-1-'09] written to Hon. G. Graham describing bitter pit and its distinction from another spot of apples

McAlpine (1909c) A letter [5-1-'09] to Secretary of the SA Dept of Agri drawing attention to the need for joint action by the Colonist Government to overcome the problem of bitter bit

McAlpine (1909d) Report [20-6-'09] to Minister Graham about the nature and knowledge of bitter bit

McAlpine D (1910a) Report of the vegetable pathologist for 1907–10. Department of Agriculture, Victoria, Government Printer, Melbourne, pp 44–61

McAlpine D (1910b) Letter [5-5-'10] sent to Minister Graham supplying his views regarding a letter from Duffus on 30th April

McAlpine D (1910c) Letter in response to E.G. Duffus's [26-9-'10] memo asking to consider leading the bitter pit investigation

McAlpine D (1911) Letter [18-1-'11] written to Dr. S.S. Cameron expressing his preference to remain within the employ of the Victorian Department of Agriculture

McKenzie T (1911) In a letter to the Victorian Minister for Agriculture [3-6-'11] claiming a Mr McIntosh had eradicated bitter pit from the Dalkeith Castle orchard in Scotland 70 year previously

Osborne J (1911) Bitter pit – its cause and possible remedies. The Fruit World of Australia. 1st Issue Jan 1911

Pascoe HT (1910a) Letter [30-3-'10] from the Minister for Agriculture South Australia, to his Victorian counterpart's letter of 20-1-'10 suggesting a scientific investigation into the cause of bitter pit, and that an application should be made to the Senates of the Nation's Universities and to Professors of biology and botany asking for advice and assistance in the matter. In a footnote added by Dr. S.S. Cameron on 5-4-'10, he acknowledged that Victoria was ready to cooperate in regard to Resolution 23 of the August 1909 conference of Agricultural Ministers

Pascoe HT (1910b) The South Australian Minister for Agriculture, Mr. Pascoe, wrote [30-4-'10] to the Victorian Minister of Agriculture, advising that the Southern States, Western Australia, South Australia, Victoria and Tasmania had agreed to cooperate in an investigation of bitter pit and that he expected NSW to also agree

Pascoe HT (1910c) Letter [2-8-'10] to his Victorian counterpart saying that as there is no plant pathologist available in Australia, they would need to obtain one from overseas

Shoobridge LM (1909) The covering letter, dated December '09, with the report of the National Fruit Growers' conference of Australasia, held in Perth from 26th to 28th October 1909. Mr Shoobridge, a leading Tasmania fruit grower, was President of the conference

Smith G (1911) Letter [3-6-'11] from Sydney University reporting that N. Cobb and he had reported their view to a meeting of the NSW Linnean Society that bitter pit was caused by the bites of small insects such as mosquitoes

Sorauer P (1909) Handbook der Pflanzenkrankheiten und Pflanzenshotes, 3rd edn. Eugen Ulmer, Stuttgart

Turner JC (1910) Report of the senior inspector, fruit exports and imports. Report of the Department of Agriculture, Victoria for the years 1907–1910. J. Kemp, Govt. Printer, Melbourne, pp 213–228, cf p 221

Williamson MR (1911) Letter [13-5-'11] from former Government inspector of orchards, Wellington, New Zealand, to Minister Graham, stating that bitter pit had been evident in New Zealand for 50 years

Wilson HJP (1911) Letter [14-1-'11] from South Australian Minister for Agriculture to his Victorian counterpart suggesting that sufficient inducement be offered to McAlpine to persuade him to accept the Commonwealth's offer

Wortmann J (1892) Uber die sogenannten "Stippen" der Äpfel. Landwirtschaftliches Jahrbuche 21:663

Chapter 8
The Poison Theory

8.1 Codlin Moth, Lead Arsenate and Bitter Pit

In spite of McAlpine's apparent reluctance, the decision was made and McAlpine was persuaded to head the investigation. Following the March meeting of State Ministers, the Prime Minister supported the proposal to establish the investigation, agreed to set up a Commonwealth Advisory Committee on Bitter Pit in accord with the general structure proposed by Wilson (1911), and to appoint Daniel McAlpine to head it.

In the meantime, the Victorian Government had awarded Professor Ewart £300 for a Research Scholarship for investigating bitter pit. This had resulted through the correspondence between Minister Graham, Daniel McAlpine and Professor Ewart during January to May of 1910. In July 1910 Dr. Jean White was appointed Research Fellow and by the end of the year she published a paper comparing the enzymes of healthy and bitter-pitted apples (White 1911b). The following March, shortly after the announcement of the recommendations to the Prime Minister to appoint McAlpine to investigate the problem, made following the Conference of Ministers, Professor Ewart claimed that Dr. White had evidence that the cause of bitter pit was lead arsenate poisoning. Lead arsenate had been introduced in 1892 as a spray to control codlin moth (Fig. 8.1), a serious pest of pome fruits. It later transpired that White had thought of this possibility only a few days before Ewart announced it and she had wanted to investigate the hypothesis for a further 12 months before making any public announcement. However, in all probability Professor Ewart leaked it as the first salvo in his attempt to prevent the investigation being led by McAlpine. Once Government Officers heard the rumour, they demanded to have the hypothesis explained to them. Ewart obliged by presenting White's hypothesis to the Royal Society of Victoria 3 weeks later in April 1911. This provoked strong reactions from orchardists and government advisors who did not agree with Ewart and White's conclusions on the grounds that bitter pit was well known in orchards which had

D.G. Parbery, *Daniel McAlpine and The Bitter Pit*,
DOI 10.1007/978-3-319-09552-3_8

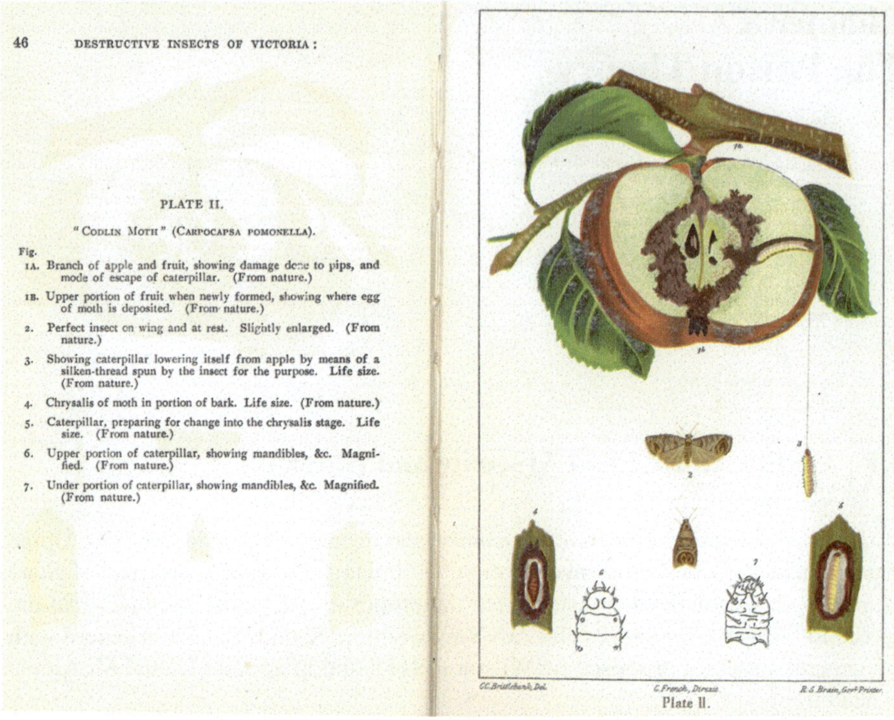
46 DESTRUCTIVE INSECTS OF VICTORIA:

PLATE II.

"CODLIN MOTH" (CARPOCAPSA POMONELLA).

Fig.

1A. Branch of apple and fruit, showing damage done to pips, and mode of escape of caterpillar. (From nature.)

1B. Upper portion of fruit when newly formed, showing where egg of moth is deposited. (From nature.)

2. Perfect insect on wing and at rest. Slightly enlarged. (From nature.)

3. Showing caterpillar lowering itself from apple by means of a silken-thread spun by the insect for the purpose. Life size. (From nature.)

4. Chrysalis of moth in portion of bark. Life size. (From nature.)

5. Caterpillar, preparing for change into the chrysalis stage. Life size. (From nature.)

6. Upper portion of caterpillar, showing mandibles, &c. Magnified. (From nature.)

7. Under portion of caterpillar, showing mandibles, &c. Magnified. (From nature.)

Fig. 8.1 Painting of the Codlin Moth from Charles French (1891) *Destructive Insects of Victoria* Vol. 1, Plate II. (Permission Department of Environment and Primary Industries, Victoria)

never been sprayed, and well before the introduction of the codlin moth and thus the use of the spray.

At the end of April McAlpine (1911b) wrote to the Hon. George Graham telling him that the Commonwealth had been advised of Ewart and White's claim, and asking that the matter be settled as soon as possible. He went on to say: "This disease has baffled the skill of the most experienced plant pathologists in Britain, America and Germany. It is a remedy that is wanted and if this has been found, the sooner the fact is known the better for the fruit industry in all States concerned." McAlpine then made an obscure reference to a recent communication from Ewart in the Press in which he says; "Since that Conference [of the Ministers of Agriculture held in March] was held, additional evidence has come to light regarding another matter which was discussed at the conference. This is a serious accusation to make against a body of Ministers accompanied by their experts and I consider that he should explain himself. The simple question that must be answered is, has a remedy for Bitter Pit of apples and pears been found – yes or no?" It seems that the accusation made against Ministers was that they were ill informed about the nature of bitter pit by advisors who didn't know what they were talking about, including of course, McAlpine.

8.2 Delay, Uncertainty and Objection

Because of the public interest raised by Ewart's assertion and the general response to it, the Federal Minister for Quarantine set up a committee to advise him on how he should proceed. Consequently, on the 29th April 1911, a report (Shoobridge 1911), signed by each member of the Committee, L.M. Shoobridge [A1], George Quinn[A1] and W.P. Norris[A1] recommended that since the investigation of the poison theory had already begun, the investigation by McAlpine should be postponed for 12 months. They also proposed that the £1,000 the Commonwealth had committed to the investigation in its first year should be redirected to support Professor Ewart and Dr. White's research. This immediately impacted on McAlpine. The Secretary of the Pharmaceutical Society of Australia contacted him wanting to know if he would be able to provide lectures for the forthcoming term. If not, the Council of the Pharmacy College needed to make alternative arrangements. McAlpine wrote to his Minister asking for his advice.

The report forwarded to the Minister for Quarantine delayed the start of the investigation, but not for 12 months. After becoming aware of the Quarantine Committee's advice, the Acting Prime Minister, W.M. Hughes, wrote to the Premiers of each State (Hughes 1911) stating that in view of Dr. White's hypothesis he considered it appropriate to allow time for her to conclude her work and have her conclusions tested before setting up another investigation. Wise as this decision may have been, it brought forth a storm of objections. On 16th May, the Acting Premier of Victoria, The Hon W.A. Watt (1911a) wrote to Hughes informing him of the outcome of a meeting of State Ministers of Agriculture held in Melbourne on 4th May to hear Dr. White's ideas on the "Poison Theory". The conclusion of that meeting was that evidence supporting the hypothesis was weak. He pointed out that Dr. White had been working on the bitter pit problem for only 9 months and that the possibility that spraying apples with lead arsenate was causing bitter pit had occurred to her only a few days before Professor Ewart announced their hypothesis at the Royal Society. Watt concluded by saying:

> Such haste in putting forward the theory without the support of field tests or indeed any other than the most elementary and inconclusive laboratory tests is not compatible with the caution usually shown by, and required of, reliable investigators, and promotes little confidence in the proposal that the wide and well considered investigation that the conference of Ministers decided on should be postponed (Watt 1911a).

In spite of objections to his decision of the 8th May, The Hon. Billy Hughes, advised the Premier of Victoria that he had accepted the advice of the Minister for Customs and would delay the start of McAlpine's investigation for 12 months. Reactions to this were mixed (Argus 1911), but objections against delay were extensive and forceful. Among the States, Tasmania and Western Australia saw no reason for delay, and only South Australia supported the Prime Minister. NSW and Victoria appear to have been ambivalent. While officers within each State's Department of Agriculture disregarded the poison theory, each Government felt it should be tested before the national investigation began. The NSW Government provided fruit from

Governments stores for testing, while the Victorian Government, having already provided funds to support Ewart's and White's research, agreed to give them access to the orchards at the School of Horticulture, Burnley, for further experimentation (Watt 1911b). Among orchardists and horticultural experts including Victorian Department of Agriculture's horticultural officers, however, there was less ambivalence. Throughout May, they opposed the poison theory on the grounds that bitter pit was common in orchards which had never been sprayed with lead arsenate. Mr. H.C.C. Anderson (1911), Victorian State Under Secretary for Agriculture presented this widely expressed assertion to the Acting Prime Minister in a letter dated 26th May 1911. Ewart received no additional funding.

A letter from P.J. Carmody (1911), Chief Inspector of Orchards, Victorian Department of Agriculture, to Dr. Cameron provides an authoritative insight into the background of the "Poison Theory". On 12th May Carmody wrote:

> Sir, I beg to draw your attention to the report in this morning's papers relative to the alleged discovery of the cause of bitter pit by Prof. Ewart and Dr Jean White. It is to be very much regretted that these co-workers have arrived at their conclusions from obviously imperfect and superficial data. It is the common practice of fruit growers not to spray young trees, as the crop would not justify the expense, and yet on these unsprayed trees with their overdeveloped fruit we find the worst attacks of Bitter Pit.
>
> Spraying with arsenical compounds has only become general during the past three or four years, whereas Bitter Pit has been rampant for many years. Trees sprayed continuously during the past year showed no sign of Bitter Pit, whereas one orchard at Talbot, though sprayed but once in the early part of the year, was the worst I have seen for many years with this disease. This was owing, not to the effect of spraying, but to the irregular development due to the improper method of irrigation. I have seen the crop on half of a Northern Spy apple [tree] badly infested with Bitter Pit while the other half was practically free from the disease, owing to the restriction of sap movement by the partial breaking down of the limb.
>
> Orchards that had never been sprayed have been seriously attacked with Bitter Pit, so I am afraid the Professor and his colleague will have to look further for causes, and it would be just as well if they would take into consideration the advisability of consulting some one who has had greater experience before publishing immature and hasty conclusions which will have the effect of interfering considerably with legitimate efforts to deal with the disease.

8.3 *Penang* – Investigative Journalist

Other interested parties protested at the proposed delay. Mr. W.H. Carne (1911), Secretary of the Gembrook South Fruit Growers' Association, wrote to Minister Graham, informing him that the Association was unanimously of the view that arsenate of lead was not the cause of bitter pit and that reports to that effect might harm the industry. He also told the Minister that the Association wished to protest against the delay in starting a thorough investigation by Mr. McAlpine. On 20th May, an article in *The Australasian* written by "Penang" (1911) stated that fruit growers were surprised by the announcement by Mr Tudor, Minister for Customs, that it was inadvisable to proceed with the bitter pit investigation until Dr. White's work was completed. "Penang" reported that the growers saw this as an attempt by the Minister to resile from the contract between the Commonwealth and States. It was pointed out that the Federal Government had offered to subsidise an investigation into the cause and control of bitter pit if the States contributed their share of expenses. This was agreed at a following conference of Ministers for Agriculture that had also selected the most skilled plant pathologist available to conduct the investigation, Mr. Daniel McAlpine. The fact was also highlighted that White and Ewart had said they would have preferred to delay putting forward their theory on the cause of the disorder until they had worked on it for a further season.

"Penang" went on to reveal that at the beginning of May, Dr. Norris, Director of Quarantine, had convened a meeting of experts from various States to discuss the investigation of bitter pit. He reported that the meeting had been held in Melbourne and that although fruit experts from neighbouring States were involved, every care was taken to maintain confidentiality to ensure Victorian experts were unaware of the meeting. None of the Victorian officers were invited to attend. When asked about it Messrs McAlpine, Carmody and Turner proved to be completely ignorant of the meeting. Whether or not Professor Ewart was aware of it is unknown. Possibly Dr. Norris wished to obtain independent advice from experts with no direct involvement in the bitter pit controversy. Later revelations, however, suggest Ewart would have known about it, which caste shadows over Dr. Norris's motives for his actions!

On 23rd May, W. Catton Grasby (1911), an agricultural expert and writer, wrote to McAlpine, possibly as a consequence of the article by Penang, and said; "Remembering what you told me about Professor Ewart when I was in Melbourne 3 years ago [1908] I have connected this supposed discovery with Ewart's antagonism to you." Grasby went on to tell McAlpine that bitter pit had been present in South Australia since about 1880 when he had seen it in his father's orchard on the variety *Cleopatra*, and when at the time there was no codlin moth, and thus no lead arsenate spray in use. He also reported that there was no codlin moth in Western Australia, that lead arsenate sprays were not used there and that although bitter pit was not a serious problem in that state, it was common enough to be noticed. On 26th May, the NSW Government also wrote to Prime Minister Hughes registering its strong disagreement that lead arsenate was involved in bitter pit (Anderson 1911).

After "Penang" blew the whistle on the secret meeting, Professor Ewart again tried to bolster the poison theory. On 29th May, *The Argus* reported his assertion that because there was no codlin moth in Western Australia, lead arsenate spray was not used, and therefore there was no bitter pit! In the same issue of *The Argus*, however, there was also a letter from Mr. W.C. Grasby' reporting that bitter pit did occur in Western Australia, thereby refuting the poison theory. Another letter in the same issue of *The Argus* highlighted the threat posed to the control of the devastating pest, codlin moth, by the stance taken by Professor Ewart. Penang's revelation prompted a lively public response. Mr. A. Renard (1911) refuted the poison theory in favour of the cause being nutrient imbalance involving potassium. On 2nd June, Dennis, Lascelles, Austin & Company wrote (1911) to Minister Graham, supported the Geelong and Western District Fruit Growers, who had passed a resolution based on their observations of many years, that there was no connection between spraying trees with lead arsenate and the occurrence of bitter pit, and asked that their information should be conveyed to the Minister. They wrote urging the Minister to keep McAlpine at work investigating the cause and control of bitter pit.

Also on 2nd June, *The Age* (Anon. 1911), *The Argus* and several local newspapers reported that a deputation of Growers had descended on Mr. Tudor in consequence of Penang's article about the secret meeting in early May. They complained that Mr. McAlpine had not been involved in the meeting and supported the widely held view that spraying with pesticides played no role in causing bitter pit. Both newspapers also revealed that the sum being considered to support the bitter pit investigations was in the order of £8,000 to £10,000 and that the Federal Government had agreed to provide half the funding if the States contribute the other half.

On 12th June, McAlpine (1911d) sent a memorandum to Dr. Cameron raising concerns about Professor Ewart's behaviour in changing earlier statements made about their poison theory. These were partly based on critical observations made by his assistant, Charles Brittlebank, who had attended Ewart's presentation to the Victorian Royal Society meeting in late April. Brittlebank was critical of several aspects of what Ewart had told his audience. He considered these were either erroneous or contrary to what White and Ewart had in the annotated manuscript. He provided McAlpine with separate notes on 12 points with the annotated typescript. Unfortunately the latter has not been found so that the significance of some of the comments is not evident. It is evident, however, that there were inconsistencies between what was written and what was said, and that changes had been made in order to circumvent critical points made against the theory. For example, although Dr. White stated that she first had the idea at the end of March that lead poisoning may have been the cause of bitter pit, a month before Professor Ewart claimed they had found the cause, Ewart let slip that in the previous October he had sprayed one portion of his own orchard with lead arsenate and left the other portion unsprayed, which was a 5 month's discrepancy between the two statements. Brittlebank also pointed out a number of statements in the document, which demonstrated that Professor Ewart had not done his homework.

For example, Ewart did not know that bitter pit occurred in Western Australia when he claimed it was because there was no need to spray from codlin moth which did not occur there. He also claimed that bitter pit first appeared in Victoria and elsewhere in 1892 after lead arsenate sprays were introduced. It was common knowledge within the horticultural industry, however, that pit was evident well before the introduction of the harlequin bug (Fig. 8.2) and codlin moth between 1870 and 1880. Also, when he was asked why pits were more concentrated at the calyx end of young fruit than near the point of attachment, Professor Ewart held up an apple by the stalk to show how the spray ran down to the calyx end on the supposition that the spraying was done when the apple was hanging down. This suggested that Professor Ewart was unaware that following petal drop the stalk remains more or less vertical with the young fruit and calyx uppermost. As the weight of the developing fruit increases, it pulls the fruit down so that the calyx ends up underneath!

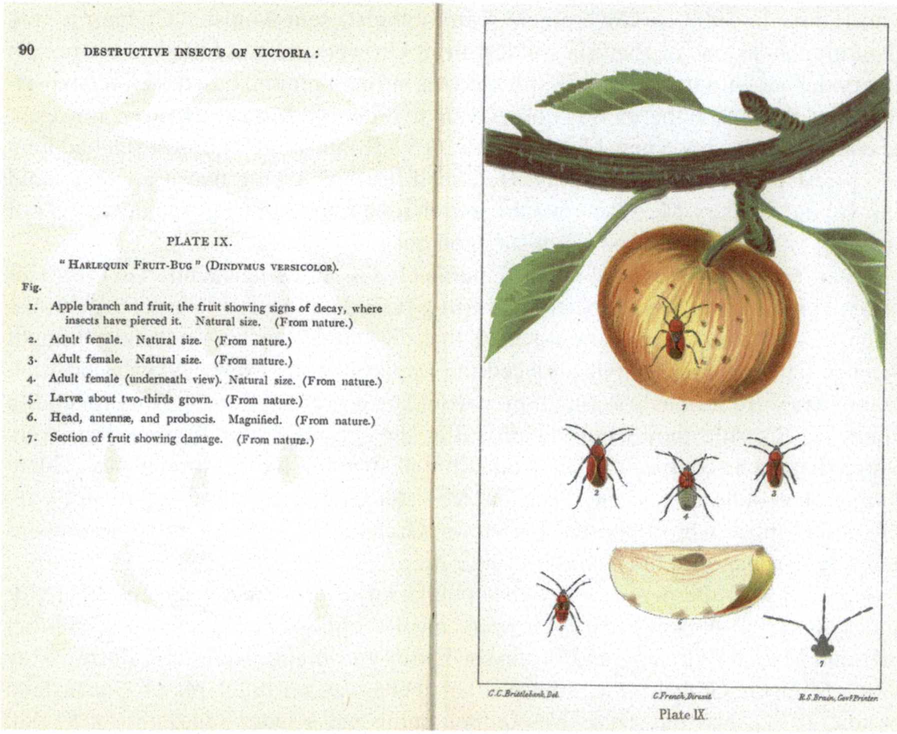
90 DESTRUCTIVE INSECTS OF VICTORIA:

PLATE IX.

"HARLEQUIN FRUIT-BUG" (DINDYMUS VERSICOLOR).

Fig.

1. Apple branch and fruit, the fruit showing signs of decay, where insects have pierced it. Natural size. (From nature.)
2. Adult female. Natural size. (From nature.)
3. Adult female. Natural size. (From nature.)
4. Adult female (underneath view). Natural size. (From nature.)
5. Larvæ about two-thirds grown. (From nature.)
6. Head, antennæ, and proboscis. Magnified. (From nature.)
7. Section of fruit showing damage. (From nature.)

Fig. 8.2 Painting of the Harlequin Bug from Charles French (1891) *Destructive Insects of Victoria* Vol. 1, Plate IX. (Permission Department of Environment and Primary Industries, Victoria)

McAlpine ended his memorandum (1911d) implying that Ewart had removed a large pitted apple from an exhibit intended to show the difference in pit susceptibility between large and small apples. He said he had exhibited a pitted apple of *Munro's Favourite* weighing 32 onz [0.91 Kg] at the Horticultural Society's Show held in April. It was the heaviest apple he had ever seen and it was the only pitted apple at the show. He believed it was stolen and his attempts to recover it failed. However, Dr. Jean White told him that Ewart had given her a very large apple, which was the only pitted apple of *Munroe's Favourite* she had ever seen, and that she used some of it for her experiments.

8.4 The Minister and the Poison Theory

Soon after the Poison Theory was announced, McAlpine (1911a) wrote to Dr. Cameron informing him that lead arsenate spray was not used in Victorian orchards prior to 1892, and 2 weeks later he offered to provide Cameron with a bitter-pitted apple from an unsprayed tree. On 12th May, Charles French (1911), who retired in 1908 as Government Entomologist, sent Minister Graham a note claiming that as one of Victoria's oldest Fruit Growers, he had followed the poison theory debate with interest, and expressed his strong opinion, based on long experience, that the poison theory was not relevant to bitter pit. More interesting, however, he claimed it was not a new idea since Mr. G.W. Robinson of Narrewarren had long ago proposed the same possibility. He ended his note saying that if anyone could unravel the mystery McAlpine would, and that he hope to live to see the effects of bitter pit very much reduced if not stamped out.

Later in May, McAlpine also sent a memo (1911c) drawing attention to Professor Ewart's claim to the Royal Society that Bitter pit did not occur in Western Australia, when in fact it did. McAlpine asserted that the whole business not only brought Science into contempt, but also placed the Department in a very awkward position as growers were seriously asking if they would be prevented from spaying for codlin moth? As a result, they would again suffer the serious losses from codlin moth, which they incurred prior to the introduction of spaying. By this time it was evident from correspondence and the press that Minister Graham was losing patience with Ewart and those who supported his views. Graham himself had an apple orchard and was familiar with the history of bitter pit.

On 16th May the *Argos Newspaper* published several articles (Argus 1911) for and against the poison theory. Growers mostly oppose it. Significantly, conflict between Minister Graham and Professor Ewart was highlighted. Also during May 1911, Minister Graham was advised that there was no bitter pit in Queensland (Anon. 1911), and the *Argus Newspaper* published a letter suggesting a $1,000 reward should be offered the person who discovered a cure for bitter pit (Anon. 1911). No reward was ever offered, and had it been, it would have been a long time coming!

8.5 Test Committee Fails Ewart and White

The following day, both *The Age* and *The Argus* reported that the Government had organised a test of Professor Ewart and Dr. White's poison theory. The Committee of three appointed to oversee the test include Mr. C.A. Nobelius[A1] (Nurseryman from Emerald), Mr W. Laidlaw[A1] (Plant Pathologist/Horticulturalist, Victorian Department of Agriculture) and Mr. R. Scott (Chemist, Victorian Department of Agriculture). Two men were appointed to supervise the collection of affected fruit from four selected orchards, Dr S.S. Cameron[A1] (Director of Agriculture, Victoria) and Mr. P.J. Carmody[A1] (Chief Orchard Supervisor, Victorian Department of Agriculture). Cameron and Carmody were also responsible for obtaining affidavits, from the four orchardist required to attest to the history of spraying the test trees in their orchards. A third group comprising Dr. Jean White[A1] (Research Fellow, Department of Botany, The University of Melbourne), Dr. W. P. Norris[A1] (Chairman, Commonwealth Director of Quarantine within the Department of Trade and Customs) and Professor Ewart[A1] (Head of the Department of Botany at the University of Melbourne and Victorian Government Botanist), was responsible for supervising the chemical testing. On the same day there was a number of letters from orchardists in *The Age* and *The Argus* stating that the authors knew that bitter pit was common in orchards, which had never been sprayed with lead arsenate or anything else.

Within a month, on 12th July 1911, *The Argus* Newspaper published the following:

> BITTER PIT
>
> **Poison Theory Discounted**
>
> The Minister for Customs (Mr. Tudor[A1]) has received from the Acting Premier (Mr. Watt) a letter on the subject of Bitter Pit setting out the results of the special investigation now concluded.
>
> 1. That the whole Committee agreed (with the exception of one apple described by Professor Ewart as doubtful) as to the existence of Bitter Pit in the apples collected.
> 2. Trees from which the apples had been collected had not been sprayed for periods of 2 to 9 years [different orchards, different periods].
> 3. That Bitter Pit was as prevalent during the present and past season on unsprayed trees.
> 4. Microscopic examination confirmed diagnoses of Bitter Pit made from observations of external symptoms.
> 5. Chemical analysis showed no trace of mineral poison.
>
> Thus the theory of Dr White and Professor Ewart is disproved.

In the same edition, it was also reported that agreement had been reached between the Commonwealth and six State Governments to set up a Commonwealth Advisory Committee on Bitter Pit and that the investigation was to be led by Mr. Daniel McAlpine on a salary of £1,000 per annum plus £1,000 per annum for experimental purposes and travelling expenses etc. for 4 years.

References

Anderson HCL (1911) Letter [26-5-'11] from Under Secretary for Agriculture, NSW, to Acting Prime Minister Hughes stating "The NSW Govt does not agree with the lead arsenate poisoning theory since many orchards in NSW, badly affected by bitter pit, have never been sprayed."

Anon. (1911) Report [2-6-'11], which appeared in *The Age* Newspaper stated: **(i)** A Deputation of Victorian Fruit Grower waited upon the Minister for Customs led by the President of their Association, Mr. Tulley. They complained that Mr. McAlpine had been displaced from those conducting the investigation into the proposition that arsenate was the cause of bitter pit. Mr. Tudor was quoted as responding with, "I do not think Mr. McAlpine ever had a part in it. I have done nothing to displace him from those conducting the inquiry." The Deputation then told him that their Association knew that bitter pit had been present long before spraying began. **(ii)** A meeting of Geelong and District Fruit Growers' Association had resolved to inform the Minister that spraying was in no way responsible for bitter pit. **(iii)** A deputation of Victorian Fruit Growers' Central Association urged Tudor to expedite the enquiry and claimed that Mr. McAlpine had been stopped in his experiments. They asked that he be reinstated. Tudor responded by saying that work had not begun and it had only been postponed. He said it could be a waste of time and money if the investigation went ahead and then the inquiry found the cause was lead arsenate poisoning

Argus Newspaper (1911) Published [16-5-'11] several articles for and against the poison theory. Growers mostly oppose it. Significantly, conflict between Minister Graham and Professor Ewart were highlighted

Carmody PJ (1911) Letter [12-5-1911] to Dr. S.S. Cameron advising him that Ewart & White's poison theory was highly unlikely

Carne WH (1911) Letter to Minister of Agriculture [15-5-'11] on behalf of the Gembrook South Fruit Grower's Association

Dennis, Lascelles, Austin & Co (1911) Letter [2-6-'11] from this firm of Wool Brokers, Stock, Station and Shipping Agents, Geelong to Hon. G. Graham, in support of the Geelong and Western District Fruit Growers' Association informing him of the outcome of the Associations meeting the day before when they resolved: **(i)** That it was the opinion of the meeting, based on practical experience, that bitter pit was not caused by lead arsenate; and **(ii)** That the meeting thanked the Minister for his interest and support, and urged him to ensure that Mr. McAlpine was kept on in his investigation of the cause

French, Charles Snr (1891) Destructive insect of Victoria, vol. 1. Department of Agriculture Victoria, Govt Printer, Melbourne

French, Charles Snr (1911) French wrote a note to Minister Graham [12-5-'11] expressing his view that there was no substance to the Poison theory, which had been proposed from another quarter some years before

Grasby WC (1911) Letter written to D. McAlpine [23-5-1911]

Hughes, The Hon. WM (1911) Letter [8-5-'11] sent to Hon. W.A. Watt, Acting Premier of Victoria

McAlpine D (1911a) Wrote [28-4-'11] to Hon. G. Graham informing him that the Commonwealth Government had been advised of Ewart & Whites poison theory of bitter pit

McAlpine D (1911b) Letter to Dr. Cameron [28-5-'11] providing a list or orchards where late in the season they could obtain apples with bitter pit from unsprayed trees for the testing of the poison theory

McAlpine D (1911c) Sent a note either to The Minister of Agriculture or Dr. Cameron [29-5-11] drawing attention to the confusion and concern among growers over whether or not they would be allowed to continue spraying and the potential damage to crops if the had to stop

McAlpine D (1911d) Sent a Memorandum [12-6-'11] to Dr. Cameron raising his concerns about the reliability of aspects of Professor Ewart's presentation of the Poison Theory of bitter pit

Penang (1911) Article [20-5-'11] by an author using the nom-de-plume *Penang* entitled "Fruit Growing – The Bitter Pit: Further Procrastination", in The Australasian Newspaper, p. 2

Renard A (1911) Letter [30-5-'11] to *The Age* newspaper stating his view that bitter pit was not due to poisoning by lead arsenate, but was due to nutrient imbalance, especially potassium. On 3-6-'11 he wrote a similar letter to Hon G. Graham

Shoobridge LM (1911) Report of 29-4-1911, sent by the Committee appointed by the Federal Minister for Quarantine advising him to delay the start of the bitter pit investigation until Ewart and White's claim that the cause of bitter pit was due to lead arsenate poisoning was fully tested

Watt WA (1911a) Letter [16-5-'11] written to Prime Minister Hughes

Watt WA (1911b) Letter [18-5-'11] informing Prime Minister of Support

White J (1911a) Bitter pit and the enzymes of apples. J Dep Agric Vic 8:805–811

White J (1911b) Bitter pit of apples. Proc R Soc Vic ns 24(1):1–9

Wilson, Hon. JP (1911) Letter [14-1-'11] from South Australian Minister for Agriculture to his Victorian counterpart suggesting that sufficient inducement be offered to McAlpine to persuade him to accept the Commonwealth's offer

Chapter 9
A National Investigation

9.1 Government in Australia

It may assist readers to understand that at the time the bitter pit investigation was being considered, there had been Federal Government in Australia for less than a decade. Following federation of the various Colonial Governments in 1901, Australia had a central Federal [Commonwealth] Government responsible for defense, foreign affairs, taxation, etc., which today is based in Canberra, and there were six State Governments. There were also the Northern Territory and the Australian Capital Territory governed directly by the Federal Government.

In 1901 an enclave of 900 square miles of land in New South Wales was set aside as the Australian Capital Territory, the future site of the Australian Federal Government. At that time, however, Canberra (not named as such until 1913) was little more than a village. Therefore, as no Commonwealth Parliament building existed, the Victorian State Government moved its operation to the Exhibition Building, opened in 1880 in Carlton, and made Victoria's Parliament Building in Spring Street available for use by the Federal Parliament. From 1901 until 1926, the Australian Federal Parliament met it Melbourne, and its various departments were also based there. Because of this, the bulk of the bitter pit correspondence was generated from, or received in, offices in Melbourne, mostly within walking distance of each other. Not being aware of this could create the impression that the bitter pit investigation was primarily a Victorian investigation, and that Australia's postal service was ahead of its time in speed of delivery! Certainly, Melbourne was McAlpine's base and was where he conducted many experiments and trials. He also established field trials and experiments in every state, visited them regularly over the first 4 years, and received assistance from cooperating horticultural officers in every state.

D.G. Parbery, *Daniel McAlpine and The Bitter Pit*,
DOI 10.1007/978-3-319-09552-3_9

9.2 Appointment of McAlpine

Meanwhile, it is evident that those interested in getting the investigation underway were making arrangements behind the scenes. On 1st July, Dr Cameron had forwarded to each State Minister an outline of the agreement between the Commonwealth, the States and Mr McAlpine. On 13th July, The Hon. James Mitchell, Minister for Agriculture and Industries for Western Australia, expressed his pleasure in endorsing the action taken and noting the salary being paid to McAlpine (Mitchell 1911). In a similar letter sent to Minister Graham the Hon. Alex Heay, Minister of Lands and Works, Hobart, also approved the action taken and pointed out that bitter pit had been evident in Tasmania since long before the introduction of codlin moth, so that in the opinion of the Tasmanian Department of Agriculture it would be a waste of time to bother with the alleged discoveries of Dr White and Professor Ewart (Heay 1911).

Even before the investigation was under way, it is evident that the administrative arrangements between the Commonwealth Department of Trade and Customs and McAlpine were unsatisfactory. He was not being informed about schedules, etc., which were essential to his own reorganisation. Throughout the investigation McAlpine voiced his annoyance at failures to keep him informed. For example, 1 week before the formal start of the investigation, McAlpine was unaware of the starting date! On 25th July he (McAlpine 1911a) wrote to Dr Cameron, telling him that as the next session of lectures was due to begin at the Victorian College of Pharmacy on 31st July, he was required to advise the President of the College whether or not he would be available to provide them. Minister Graham replied, in a letter dated 31st July, telling him that he was requested to take up his duties on 1st August! It is therefore to be hoped that Dr Cameron had been able to advise him of this verbally some days sooner! It is evident, however, that he was not kept abreast of what was being organised. On 2nd August, McAlpine (1911b) wrote to Minister Graham acknowledging his letter of 31st July, and telling him that the federal authorities had advised they would provide him with a typist to assist in documenting his work. At the end of the letter he said; 'I presume an agreement has been drawn out stating the terms and duration of the investigation.' The following day McAlpine (1911c) sent a memo to Dr Cameron:

> Since there is no representative of the Fruit Growers on the Advisory Board, might I suggest that you mention this fact to Mr Tudor and if necessary the well-known name of Mr Lang JP of Harcourt might be mentioned.
>
> Could you suggest a short pithy title for the position I have been called to fill as Investigator of Bitter Pit seems rather clumsy.

On 1st August 1911, the same day as the Commonwealth Advisory Committee on Bitter Pit was established, McAlpine was asked to put aside his preferred work as Victoria's Vegetable Pathologist and devote himself to finding the cause and control of bitter pit. He was given a free hand to conduct the investigation as he saw fit, while having to report annually to the Advisory Committee. According to McAlpine's daughter Erica Wedge (1984), her father's understanding at the time was that he was being 'lent' to the Commonwealth for the duration of the investigation. He believed he had a gentleman's agreement with the Victorian Minister for Agriculture, George Graham, that his position as Vegetable Pathologist would still be available to him at the conclusion of the investigation. It is quite possible that this was the carrot which eventually persuaded him to accept the appointment. While that may well have been the case, there is no doubt that McAlpine resigned his contract with the Victorian Department of Agriculture to take up the appointment under the Commonwealth.

On 1st August 1911, Minister Graham wrote to Acting-Premier Watt advising him that everything was in place for the investigation to begin and that the matter had reached the stage where it would be expedient for all correspondence relating to the investigation to be directed through the Department of Trade and Customs. He also asked that the department supply a secretary to serve the Advisory Committee. On 2nd August, the Under Secretary of Agriculture for NSW (Anderson 1911) wrote giving NSW's approval of the proposal received in July but suggested that as McAlpine was being paid such a high salary, he could also investigate other national disease problems as well as directing his main effort to the bitter pit investigation. Dr Cameron replied (1911) that the conference of ministers held in March had directed that the investigation would be funded for a single purpose. As the agreements embodied the directive, no changes would be made.

9.3 Commonwealth Advisory Committee on Bitter Pit

Also on 1st August 1911, the Commonwealth Advisory Committee on bitter pit, hereafter referred to as the Advisory Committee, was established. It was the conduit between McAlpine and the Federal and State Governments. Eight parties were listed in the agreement, the first being the Commonwealth Government, represented by the Minister of Trade and Customs. The 2nd to 7th parties were the six states, represented by their Ministers of Agriculture and numbered in order of their nominees as listed below in the composition of the Advisory Committee. Representatives of these seven parties constituted the Advisory Committee, while Daniel McAlpine Esq. was the 8th party. Three appointees represented the Federal Government while there was one appointee for each State. The Membership of the Advisory Committee at its first meeting on 29th and 30th October 1912, consisted of the following:

Three Commonwealth Government appointees, namely

Dr W.P. Norris[A1]	Chairman of the Committee, Commonwealth Director of Quarantine within the Department of Trade and Customs.
Professor T. Cherry[A1]	Microbiologist and Professor of Agriculture, at the University of Melbourne.
Mr J. C. Turner[A1]	Senior Inspector, fruit exports and imports, Department of Agriculture, Victoria.

Six State representatives, namely

G.P. Darnell-Smith[A1]	Representing New South Wales.
W. Laidlaw[A1]	Representing Victoria.
H. Tryon[A1]	Representing Queensland.
G. Quinn[A1],	Representing South Australia.
P.J. Carmody[A1]	Representing Western Australia.
A.H. Benson[A1]	Representing Tasmania.

The only changes to the composition of the Advisory Committee over the duration of the investigation were:

Dr J.H.L. Cumpston[A1]	Replaced Dr Norris as Chairman of the Committee in May 1913.
Mr E. Meeking	Replaced Mr Turner in 1916.

Two days after he took up his appointment, McAlpine (1911c) again wrote to Dr Cameron expressing his disappointment and possible concern that an orchardist was not appointed to the Advisory Committee, and suggested Mr Lang JP of Harcourt. However, membership was confirmed and no change was made.

9.4 Travel, Distance and Time

All meetings of the Advisory Committee were held at the Department of Customs and Trade in Melbourne but, because of travel constraints, Mr Benson attended only two meetings while Mr Tryon attended none. The possible reason for this, apart from limited apple growing in Queensland, was that the distance Tryon had to travel from Brisbane to Melbourne was a little over 1,250 miles [2,000 km] by rail and took at least 70 h involving two breaks of gauge. The distance by sea was ca 1,150 miles [1,800 km] and could take up to a week depending on weather conditions, the number of ports of call, and the speed of the vessel, which was usually about 10 knots. As ships had no stabilizers, were between 500 and 1,000 t, seas were frequently rough and many passengers were seasick, travellers frequently chose alternatives if available.

It is worth noting that during the first 4 years of the investigation, McAlpine visited Queensland at least twice in 4 years and each other State at least annually. His travel was by rail and then by horse drawn conveyance from the railhead nearest to his destination. The rail journey from Melbourne to Perth was a total of 2,070 miles [3,451 km]. He would usually stop off at Adelaide, 442 miles from Melbourne and then endure the next 1,628 miles involving three changes of gauge and taking at least 50 h. Visits to Tasmania required an 18-h voyage from Port Melbourne to Devonport, then a 6-h train journey to Hobart. It is therefore not surprising that Benson travelled from Hobart for only two meetings, and that P.J. Carmody, who represented Western Australia on the Advisory Committee, was based in Melbourne.

References

Anderson HCL (1911) Letter [2-8-'11] from Under Secretary for Agriculture, NSW, to the Secretary, Department of Agriculture, Victoria, suggesting that as McAlpine's salary was generous he might investigate other diseases of national importance as well as bitter pit

Cameron SS (1911) Letter [9-8-'11] written to H.C.L. Anderson, in response to his letter of 2nd August stating that the terms of McAlpine's appointment immutable

Heay, Hon. Alec (1911) Letter [17-7-'11] written to his counterpart Hon. G. Graham re the proposed establishment of the BP investigation urging them to get started, and not waste time over the White/Ewart alleged discovery as bitter pit was present in Tasmania long before the use of lead arsenate sprays began

McAlpine D (1911a) McAlpine wrote [25-7-'11] to Dr. Cameron, advising that the College of Pharmacy began second term on 31st July and needed to know if he were availability. He asked to be told when the investigation would begin

McAlpine D (1911b) Letter [2-8-'11] responding to Graham's letter of 31st July, requesting him to begin the investigation on 1st August. McAlpine said he had done so, referred to the Commonwealth Government providing him with a typist and saying he had still not seen the terms of the agreement

McAlpine D (1911c) McAlpine sent a memo [3-8-'11] to Dr. Cameron drawing attention to the lack of a fruit grower on the Advisory Committee and suggesting Mr. Tudor might consider Mr. Lang JP of Harcourt. He also asked Mr. Cameron to suggest a more pithy title for his new position other than "Investigator of bitter Pit"!

Mitchell, Hon. J (1911) Letter [13-7-'11] sent from W.A. Minister of Agriculture and Industries to Victorian counterpart endorsing the establishment of the BP investigation and noting McAlpine's appointment, in response to advice on 11th

Wedge E (1984) Letters [4th to 7th May-1984] to D.G. Parbery, La Trobe Collection

Chapter 10
The First Year 1911–1912

10.1 Establishing Starting Points

Towards the end of each year, McAlpine submitted a Progress Report to the Advisory Committee describing experiments and tests conducted, and the conclusions drawn. Following discussion of the annual reports with McAlpine, the committee then furnished its own report to the Prime Minister's Office and each State Premier. As the investigation progressed, the Advisory Committee began to make suggestions on changes or improvements that might be made to some field trials, and in some case strongly urged McAlpine to adopt them. Although its terms of reference gave the Advisory Committee no authority to direct McAlpine, he accepted some as suggestions while objecting to some and ignoring others, even when they were re-emphasised in a following report.

In the introduction to his first report, prior to explaining the objective of the various trials and experiments, McAlpine said '*It must never be forgotten that the final test of the results obtained must be their actual value to the practical man*!' In doing so, he plainly stated his primary objective was to reduce the problem.

10.2 The Australian Apple Industry 1900–1911

At the outset of the investigation he established benchmarks against which he might have expected his masters to gauge his success. He provided statistics on the production of apples in Australia over the 3 years leading up to the beginning of the investigation. These provided a baseline from which to evaluate the scale of the problem and allowed evaluation of benefits the investigation might bring to the horticultural industry.

The data in Table 10.1 demonstrate that the two leading apple-producing states were Victoria and Tasmania which produced 34.84 % and 33.69 % of the Australian

D.G. Parbery, *Daniel McAlpine and The Bitter Pit*,
DOI 10.1007/978-3-319-09552-3_10

Table 10.1 Production of apples in the commonwealth of Australia

Years				
States	1908–1909	1909–1910	1910–1911	Totals
Qld	31,121	29,662	25,410	86,193
NSW	385,649	474,838	596,561	1,457,048
Vic.	1,241,826	1,121,702	1,667,271	4,030,799
SA	398,812	557,130	476,904	1,432,846
WA	186,321	217,533	261,563	665,417
Tas	1,070,546	1,480,107	1,347,952	3,898,605
TOTALS	**3,314,275**	**3,880,972**	**4,375,661**	**11,570,908**

From the Department of Agriculture Report Victoria, 1907–1910, p. 221
Quantities are given in bushels, 1 bushel of apples = 42 lbs [19 kg]

Table 10.2 Total apple exports 1909–1911[a]

	Qld	NSW	Vic	SA	WA	Tas	Totals
1	216	63,097	707,137	267,454	25,837	2,109,390	**3,173,131**
2	0.25 %	4.33 %	17.54 %	18.67 %	3.88 %	54.11 %	
3	0.01 %	1.99 %	22.39 %	8.35 %	0.81 %	66.50 %	**99.97 %**[a]

From the Department of Agriculture Report Victoria, 1907–1910, p. 221
[a]A small proportion, 0.03 %, of the nation's apple crop came from the Federal Territories and is discounted
1 = bushels exported; 2 = the proportion of each state's production which was exported; 3 = each state's proportion of Australian exports

apple crop respectively and close to 69 % between them, New South Wales and South Australia produced 12.56 % and 12.38 % respectively a little over 24 % between them, and Western Australia 6.75 %, Queensland 0.75 % and the territories 0.03 % the remainder.

The volumes and proportions of apples exported overseas are given in Table 10.2. The outstanding exporter was Tasmania while Queensland exported very few and possibly only interstate. From these data it can be seen that between them, Tasmania and Victoria together produced 89 % of the apples exported from Australia, NSW plus South Australia 10 %, Western Australia almost 0.8 % and Queensland less than 0.01 %. Turner (1910) listed Victorian apple exports for the 10 years 1900–1909 and also drew attention to the increasing prevalence of the development of bitter pit in consignments during shipment, even though all fruit had been free of it when shipped (Fig. 10.1).

In this first part of the investigation, McAlpine (1912a) emphasised what he saw as the fundamental principle of plant pathology, that the normal structure and functions of the host part or organ should be determined as far as possible to allow the abnormal to be understood, and he stated his intention to do that. He was familiar with Sorauer's texts and had done his best to bring his knowledge of plant physiology up to date by reading Ewart's (1900–1906) translation of Professor Pfeffer's three-volume text on plant physiology. In addition, he reviewed 98 publications

Fig. 10.1 Painting of bitter pitted apples by C. C. Brittlebank (Frontice Piece of McAlpine's Second Report Progress Report (1913). Courtesy of the Royal Botanic Gardens Library, Melbourne)

dealing with the problem of bitter pit. These mainly explored ideas about its cause. In the summary to his report he listed seven factors that he considered contributed to the onset of bitter pit. These included: (i) intermittent weather conditions when fruit was at its critical stage of development; (ii) the amount and rate of transpiration; (iii) sudden checking of transpiration at night when roots were still active due to heat in the soil; (iv) failure especially in the supply of water to the periphery of fruit followed by spasmodic and irregular recovery; (v) inequality of growth of the vascular system so that nutrients were not evenly distributed; (vi) fluctuations in temperature when fruit is in store, and (vii) the nature of the variety. In his first report he published detailed photographs of the vascular systems of both apples (Fig. 10.2) and pears (Fig. 10.3), that were later called in to question!

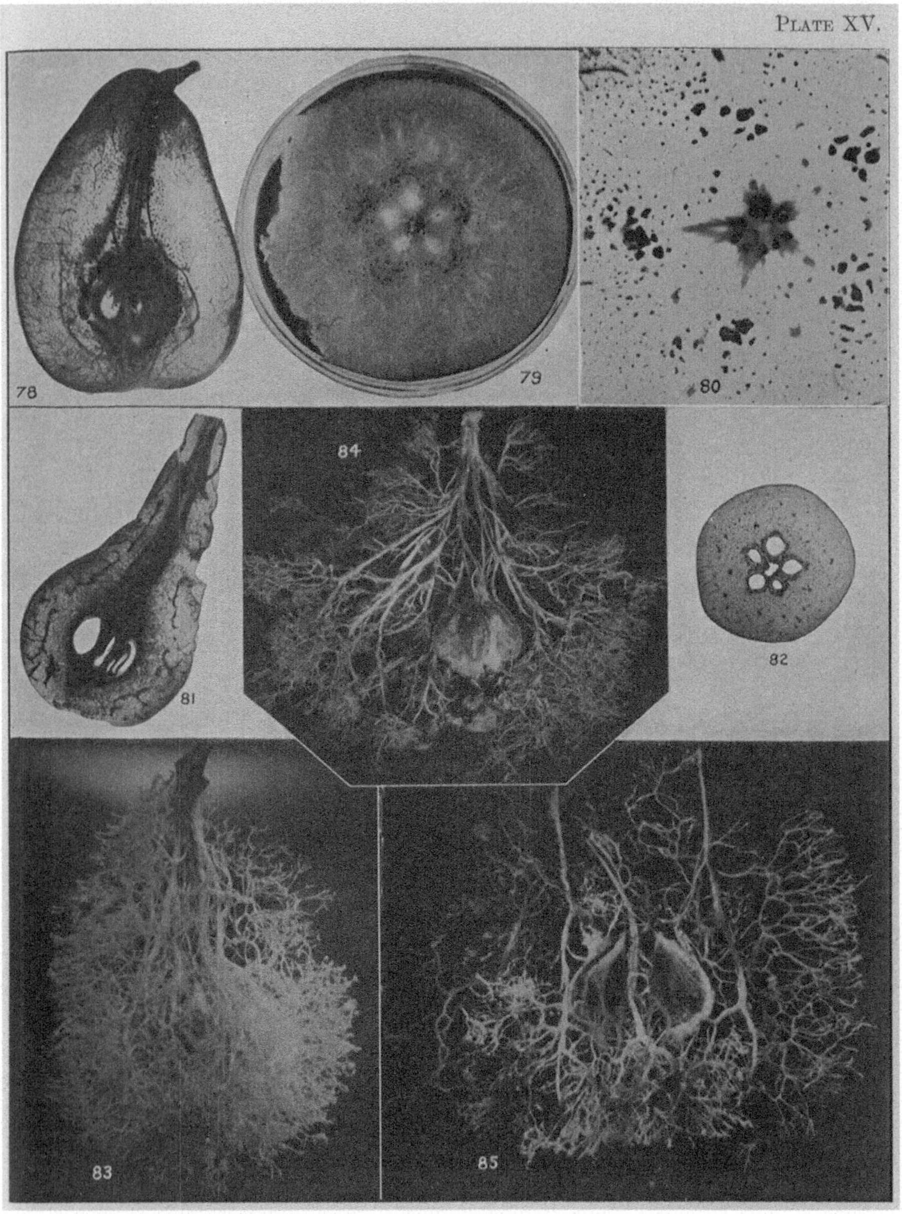

Fig. 10.2 Vascular systems of a pear showing stone cells stippled (Fig. 78) (From McAlpine's First Progress Report (1912a) Plate XV. Courtesy of the Royal Botanic Gardens Library)

McAlpine's studies of the pears and apples were just as meticulous as his studies of fungi and were as accurate as his mycological studies but have attracted less attention. He managed to erode away the flesh of the fruit he studies, leaving the vascular

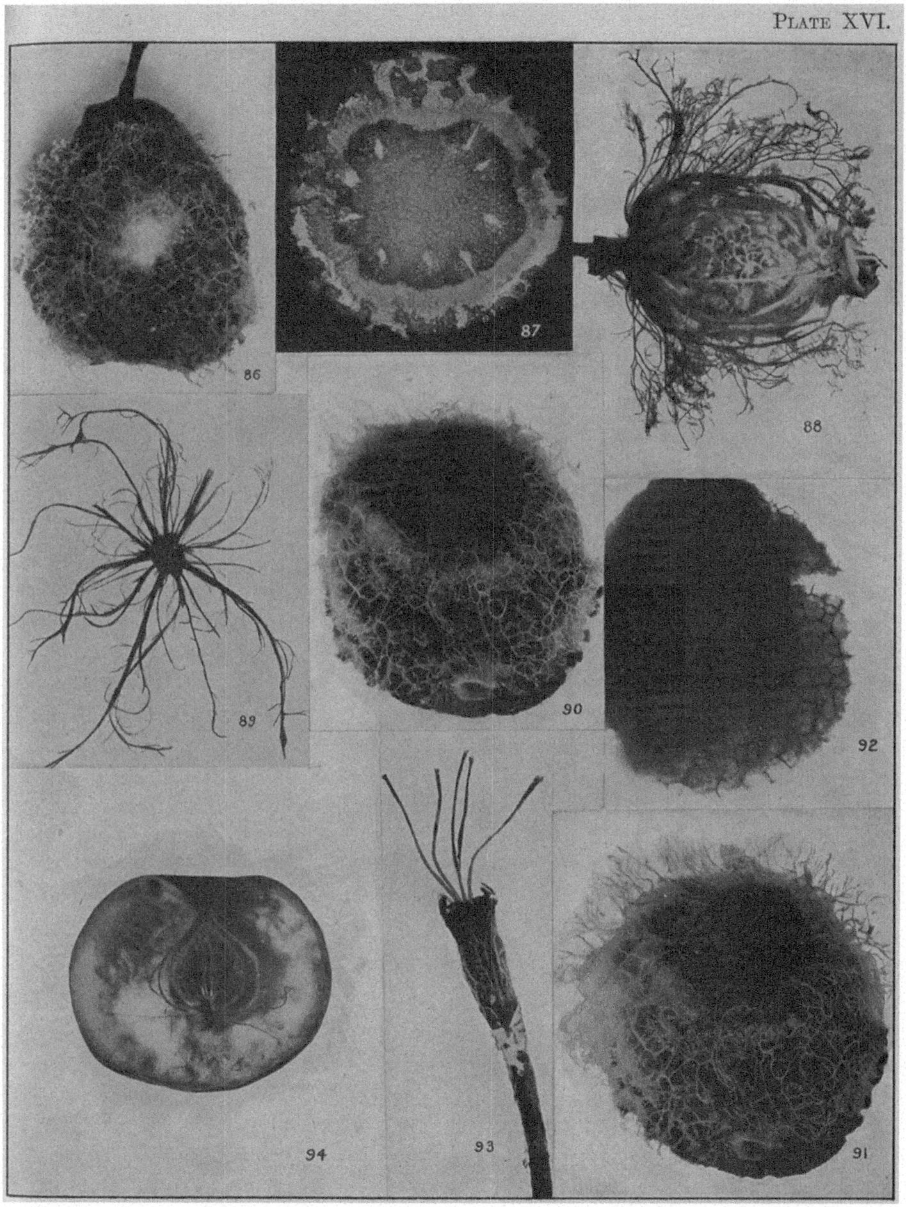

Fig. 10.3 Vascular systems of pear (figs 86 and 87) and apple (figs 88 to 94) (From McAlpine's First Progress Report (1912a) Plate XVI. Courtesy of the Royal Botanic Gardens Library)

systems exposed and with clusters of dead parenchyma cells at vein endings under the skin in of pear (Fig. 10.2, his Plate XV Fig. 83) and apple (Fig. 10.3 his Plate XVI Figs. 88 and 90).

In addition to tracing the paths of the ten main vascular bundles from the stalk to the core of apples and pears (Fig. 10.3 his Plate XV, Figs. 78 and 79) he indicated the presence of stone cells though out the flesh but concentrated around the core. These of course account for the grittiness of many older varieties of pears and especially of quinces.

He was aware, from his own observations and those of horticultural colleagues such as Carmody (1910) and Turner (1910), that some orchard practices led to greater incidences of pitting than others. These included over-pruning, the overuse of certain manures, and over-watering leading to waterlogging. Added to these were weather patterns which resulted in severe changes in growth rates, differences in varietal susceptibility to bitter pit, and interactions of all these with climatic and edaphic factors. In consequence, the first 12 months of the investigation was spent in planning and establishing experiments. As a courtesy, a week after his appointment McAlpine (1911) wrote to Professor Ewart as his colleague and Government Botanist, outlining proposals for his manure and pruning experiments. If he expected some feedback, no evidence of it exists. Orchardists were identified who could provide sites of potential benefit to research on the problem, and by meeting horticultural and agricultural officers in the various states with whom he would be working. At the end of the year McAlpine (1912c) wrote to Dr Cameron suggesting that he obtain Australian rainfall maps from Mr Hunt, the Federal Meteorologist, in order to superimpose apple-growing areas on them so as to assess if there was a relationship between rainfall patterns and the incidence of bitter pit. He also requested Cameron to ask Mr Carmody to obtain the assistance of the Victorian Orchard Inspectors to compose the Victorian maps. It is probable that he also wrote to the other State Directors to ask their orchard inspectors to provide the same service.

10.3 Advisory Committee's First Report

The investigation did not begin until 2nd August 1911, so McAlpine's *First Progress Report* (McAlpine 1912a) was submitted to the Advisory Committee in September or early October 1912. Its first meeting was held at the office of the Director of Quarantine, Spring Street, Melbourne on 29th and 30th October 1912 (Norris 1912b). All members of the Advisory Committee were present except Mr Tryon. Each had been sent a draft copy of McAlpine's report in advance of the meeting (Norris 1912a). It was discussed within the committee on the first day of the meeting, and with McAlpine at the meeting on the second day. Special matters dealt with or referred to in McAlpine's report were discussed, and proposals for the elaboration of potentially beneficial initiatives were formalised into resolutions (Norris 1912c) that were distributed to the offices of the Prime Minister and each State Premier. Each State Director of Agriculture then distributed the resolutions to their officers involved in the investigation.

The resolutions from this first meeting provide an insight into the structure and goals of the investigation, so they are listed here as adopted by the committee.

Resolutions

1. Chemical analysis of immune and especially susceptible varieties of apples:

   (a) Chemical analysis desirable at different stages of their development throughout the coming season.
   (b) Advantage would be gained if the co-operation of the Commonwealth and States could extend to having such analyses conducted in the various Commonwealth and State laboratories.
   (c) That in respect of the enzymes and other organic compounds present at various stages of development, the services of specially qualified biochemists should be obtained to carry out the analyses. Professor Cherry, Mr Laidlaw and Dr Norris constituted a sub-committee responsible to advise the Minister about who should be entrusted with the analyses and how much they should be paid.

2. Cross-Breeding of Varieties of Apples:
   That representations be made to the State Governments regarding McAlpine's proposal (Report p. 71) for cross breeding commercial varieties of apples in order to test the potential benefits of cross breeding in modifying susceptibility to bitter pit. The determination of the natural history of the susceptibility to bitter pit of various combinations of stock and scion to be recorded, and special attention to be directed to the crossing of relatively unsusceptible varieties.
   It is recognised that this phase of the investigation cannot be completed within the period of the contract with Mr McAlpine.

3. The Assistance of State Field Officers in making special observations:
   That it is desirable that the services of State Field Officers engaged in horticulture be utilised as far as is practicable in observing and recording specified phenomena in connection with the development of bitter pit and the effect of prearranged factors, and that Mr McAlpine be asked to formulate a schedule of questions and observations which he considers essential.

4. Number of trees forming subject of experiment and record of data:
   That the number of trees in each experiment should be as numerous as possible and where less than three trees make up any group, they are to be recorded separately.

5. The effect of irrigation on the incidence of bitter pit:
   That the irrigation experiments should be rigidly controlled and that, *inter alia*, records embracing the following points should be kept:

   (a) The quantity of water used on each occasion;
   (b) The period over which the irrigation extended;
   (c) Daily records of rainfall during the season or seasons during which observations are made;
   (d) The precise character of the soil irrigated in each case;
   (e) Temperature condition including wet and dry bulb records in the district where the records are made.

Unless there is a Meteorological Station in the vicinity, arrangements should be made for the installation of a rain gauge and thermometer for keeping records.

6. The Effect of Close Shelter on Transpiration:
   That special investigation be made as to the apparent influence if any on the development of bitter pit of close shelter from wind and that the State Horticultural Authorities be asked to report as to the existence of any specially sheltered or protected orchards or parts of orchards where exposure to strong winds is relatively slight;

7. Observations on the Effects of Cold Storage:

   (a) That experiments on the effect of refrigeration on retarding or preventing the development of bitter pit be extended;
   (b) That it would be highly advantageous if each apple exporting State made it a regular practice to place in refrigerated storage in the State at a temperature of 30–32° F sample cases of apples from each shipment;
   (c) That samples of the apples stored locally and of those shipped to London be examined and compared as to the prevalence of bitter pit;
   (d) That on arrival of any shipment to London or any other point of observation samples of the refrigerated apples be removed from cold storage and exposed for 1 month to ambient conditions expected in the hands of retailers, repeated observations being made during the whole period as to the development of bitter pit.

8. Relation of pruning to regular fruiting:
   That it is specially desirable that pruning experiments directed to securing regularity of fruiting and the avoidance of great disparity in the crop in different seasons be continued.

This first report of the Advisory Committee carried the weight of the Federal Government authority behind it. While it formally encouraged strong support for McAlpine, which he had already won from Horticultural Officers in all states, it included a set of resolutions couched in terms directing McAlpine to comply with them. It is noteworthy that the Advisory Committee did not have the authority to direct McAlpine, and the resolutions were based on the very points that McAlpine had listed as the starting points for the investigation (McAlpine 1912a). Norris (1912c) then sent a covering letter with a copy of the resolutions to each member of the Advisory Committee telling them that the resolutions were being sent to McAlpine directing him to put them into effect. He also said he would advise the Victorian Minister for Agriculture to request his counterparts in each state to instruct their officers to provide Mr McAlpine with the assistance he needed. The Victorian Minister for Agriculture did so in mid December (Graham 1912). Over the following weeks, replies were received from McAlpine and each State Minister of Agriculture confirming their willingness to cooperate fully. In fact, however, Norris was stepping outside his Committee's brief. Each State's Department of Agriculture knew what its responsibilities were under the Bitter Pit agreement, and McAlpine

Table 10.3 Accounts for the investigation for the financial year 1st July 1912 to 30th June 1913

Items	Cost in £	Cost in AU$ (2012)
Mr. D. McAlpine's stipend	£1,000	$2,000.00
Fares	£72/18/06	$145.85
Travel allowance	£86/8/4	$172.84
Printing the report	£259/4/3	$518.42
Printing blocks etc.	£151/13/11	$303.39
Services of a temporary analyst	£28/16	$57.60
Freight, apples, fruit cases, labour, etc.	£24/1/2	$48.12
TOTAL	**£1,623/2/2**	**$3,246.22**
Commonwealth share	£811/10/8	$1,623.08
States' share	£811/10/06	$1,623.05
Each state's share	**£135/05/01**	**$270.51**

had already referred in this First Progress Report to the unstinting cooperation he was receiving, and he drew this to the attention of Minister Graham (1912b).

The letter from the Victorian Minister for Agriculture to his counterparts in the other states, however, may well have reminded the State Governments of their commitments to provide McAlpine with access to their facilities and staff in order to carry out his research program. McAlpine lost no time in framing a questionnaire in response to Recommendation No. 3, and soon distributed it to orchardists in order to collect data he needed.

10.4 The First Year's Accounts

Another insight into the investigation may be gained from noting the items included in the accounts of the bitter pit investigation for the first full year, July 1st 1912 to 30th June 1913. The following is a compilation of two requests from the Prime Minister's Office sent to each State Premier. The first covered 9 months from July 1st 1912 to 31st March 1913 (Johnson 1913), and the second the 3 months from 1st April 1913 to 30th June 1913 (Cook 1913) (Table 10.3).

There were also administrative costs paid to the Department of Customs and Quarantine, including the salary of the typist appointed to support McAlpine.

References

Carmody PJ (1910) Report of the Chief Inspector of Orchards. Report of the Department of Agriculture, Victoria for the years 1907–1910. J. Kemp, Government Printer, pp 208–212

Cook J (1913) A letter sent from the Prime Minster's Office [12-8-'13] presenting the BP Investigation accounts from first of April to 30th June 1913

Ewart AJ (1900–1906) English translation of Professor W. Pfeffer's The physiology of plants, Volumes 1 to 3

Graham, Hon. G (1912) Letter to his counterparts in the other States [12-12-'12] forwarding a copy of the Advisory Committee's resolutions, acknowledging their assistance to McAlpine over the preceding 17 months and asking for their cooperation with McAlpine in carrying out the resolution

Johnson S (1913) Bitter pit account sent to each State from the Prime Minister's Officer for the period 1st July 1912 to 31st March 1913

McAlpine D (1911) He wrote to the Government Botanist [8-8-'11] outlining proposed pruning and manurial experiments

McAlpine D (1912a) Bitter pit investigations. First progress report – the past history and present position of the bitter pit question. Government Printer, Melbourne, 197 pp

McAlpine D (1912b) Letter written to Hon. Geo. Graham [9-12-'12] informing him that Dr. Norris was acting outside his brief by interfering in administrative arrangements and cutting across arrangements already made by McAlpine for chemical analyses to be made at the Federal Chemistry Laboratories

McAlpine D (1912c) Letter written [31-12-1912] to Dr. Cameron regarding proposal to superimpose apple growing areas on Australian rainfall maps supplied by the Federal Bureau of Meteorology

Norris WP (1912a) Letter [24-9-12] written to W. Laidlaw, and other members of the Advisory Committee, with copies of the First Progress Report and foreshadowing a date for the first Advisory Committee meeting

Norris WP (1912b) Letter [17-10-12] written to members of the Advisory Committee advising date time, and venue of the first Advisory Committee meeting

Norris WP (1912c) Memorandum [19-11-12] sent by the Secretary to Members of the Advisory Committee with amended copies of the Resolution from their first meeting

Turner JC (1910) Report of the Senior Inspector, Fruit Exports and Imports. Report of the Department of Agriculture, Victoria for the years 1907–1910. J. Kemp, Government Printer, pp 213–228, cf p 221

Chapter 11
Serious Stumbling Blocks

11.1 The Irony of Preparation

McAlpine's experience in the mid 1870s in Huxley's laboratory dissecting specimens then illustrating and labelling what he found combined with his subsequent study of the literature and a wide range of biological specimens equipped him to make detailed studies of the anatomy of pome fruit. It is evident that prior to committing himself to the bitter pit investigation he began familiarising himself with the anatomy of apples pears (Figs. 10.2, 10.3) and quinces (McAlpine 1912d). As he applied himself to defining the morbid anatomy of bitter pitted fruit, he drew heavily on the information gleaned during 1910 and 1911 (Figs. 11.1 and 11.2). Transverse median section of apples (Fig. 11.1) show the ten vascular bundles positioned around the carpels from which branches radiate out towards the surface of the fruit as well as producing vascular traces supplying the developing seed. Another aspect of the system is demonstrated in the longitudinal median sections and the exposed vascular system left by the dissolution of the flesh (Fig. 11.2). It was through these examinations that he saw that the initiation of the pits was death of parenchyma cells surrounding vein endings. This confirmed other findings and supported propositions about the cause of bitter pit. It also provided Professor Ewart with a basis to accuse him of plagiarism!

11.2 Ewart Corrupts Advisory Committee

While McAlpine agreed to put the recommendations of the first meeting of the Advisory Committee into effect without reacting adversely to them, by early November 1912 he had taken serious exception to the unprofessional behaviour of the Chairman of the Advisory Committee, Dr Norris and Dr Cherry, who it appears were manipulated into making unfounded criticism of him during the Advisory

D.G. Parbery, *Daniel McAlpine and The Bitter Pit*,
DOI 10.1007/978-3-319-09552-3_11

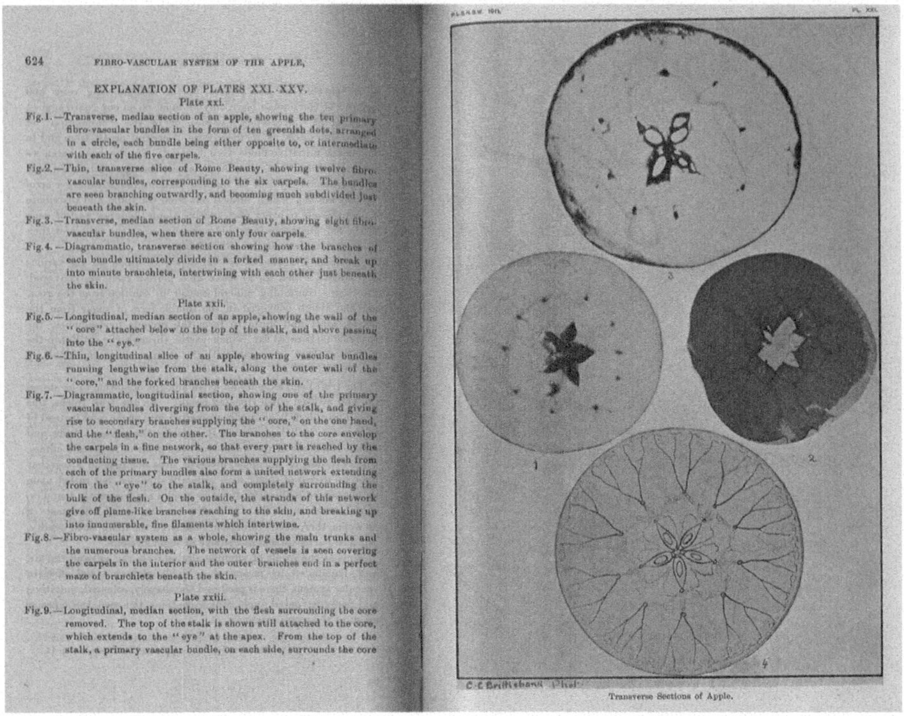
624 FIBRO-VASCULAR SYSTEM OF THE APPLE,

EXPLANATION OF PLATES XXI.-XXV.

Plate xxi.

Fig.1.—Transverse, median section of an apple, showing the ten primary fibro-vascular bundles in the form of ten greenish dots, arranged in a circle, each bundle being either opposite to, or intermediate with each of the five carpels.

Fig.2.—Thin, transverse slice of Rome Beauty, showing twelve fibro-vascular bundles, corresponding to the six carpels. The bundles are seen branching outwardly, and becoming much subdivided just beneath the skin.

Fig.3.—Transverse, median section of Rome Beauty, showing eight fibro-vascular bundles, when there are only four carpels.

Fig.4.—Diagrammatic, transverse section showing how the branches of each bundle ultimately divide in a forked manner, and break up into minute branchlets, intertwining with each other just beneath the skin.

Plate xxii.

Fig.5.—Longitudinal, median section of an apple, showing the wall of the "core" attached below to the top of the stalk, and above passing into the "eye."

Fig.6.—Thin, longitudinal slice of an apple, showing vascular bundles running lengthwise from the stalk, along the outer wall of the "core," and the forked branches beneath the skin.

Fig.7.—Diagrammatic, longitudinal section, showing one of the primary vascular bundles diverging from the top of the stalk, and giving rise to secondary branches supplying the "core," on the one hand, and the "flesh," on the other. The branches to the core envelop the carpels in a fine network, so that every part is reached by the conducting tissue. The various branches supplying the flesh from each of the primary bundles also form a united network extending from the "eye" to the stalk, and completely surrounding the bulk of the flesh. On the outside, the strands of this network give off plume-like branches reaching to the skin, and breaking up into innumerable, fine filaments which intertwine.

Fig.8.—Fibro-vascular system as a whole, showing the main trunks and the numerous branches. The network of vessels is seen covering the carpels in the interior and the outer branches end in a perfect maze of branchlets beneath the skin.

Plate xxiii.

Fig.9.—Longitudinal, median section, with the flesh surrounding the core removed. The top of the stalk is shown still attached to the core, which extends to the "eye" at the apex. From the top of the stalk, a primary vascular bundle, on each side, surrounds the core

Fig. 11.1 Transverse sections showing vascular system of the apple (McAlpine (1911a), Plate XXI. Courtesy of the Royal Botanic Gardens Library)

Committee meeting on 30th October. McAlpine was so upset that for the only time in all his correspondence, he forgot to date the letter except for '1912'! As this may well be the key to so much that followed, the letter is given here in full (McAlpine 1912a):

'The Hon. Geo. Graham MLA – Minister of Agriculture (Vic.)

Sir, My First Progress Report has now been issued and you will see from it that substantial progress has been made in the investigation of Bitter Pit. This advance in our knowledge, however, has not been agreeable to certain gentlemen who have tried to minimise it as much as possible and I am sorry to say that this attitude taken up in the first instance by Professor Ewart has been followed by Drs Norris and Cherry, who represent the Federal Government on the Committee.

1. In connection with the investigation, I made some important discoveries relating to the structure of the apple and pear and while these are described and fully illustrated in my Report (Figs. 10.2 and 10.3), they were also published by the *Linnaean Society of New South Wales* (McAlpine 1911a, b), which is one of the leading scientific Societies of Australia and of which I am a Corresponding Member. I mention this to show that while the attack was made on these published papers by Professor Ewart, it was simply because he could not publicly refer to the Report, which had been placed in the hands of Dr Norris in MS [in manuscript]. The Secretary of the Linnaean Society of New South Wales wrote to me as follows under date 13th September 1912:

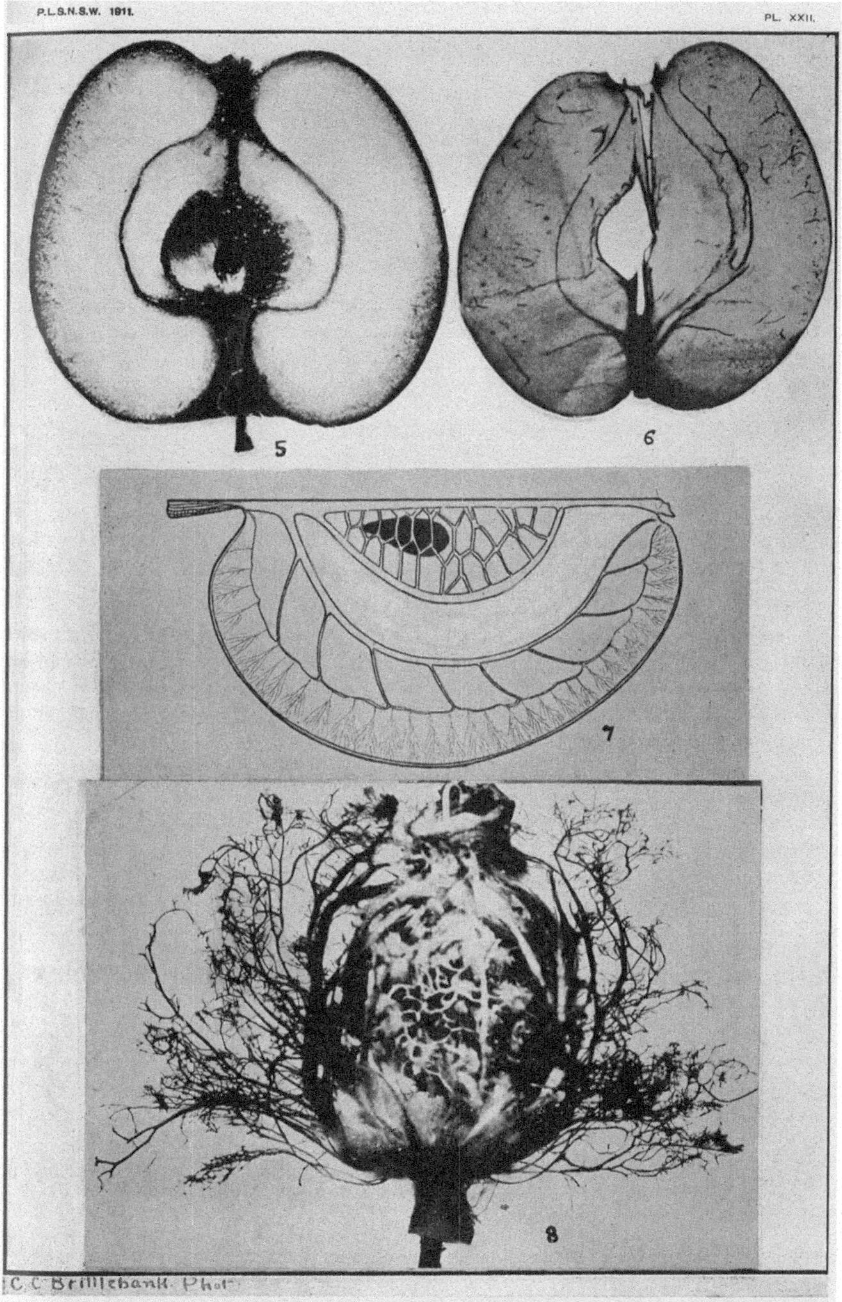

Longitudinal Sections, and Vascular System of Apple.

Fig. 11.2 Longitudinal sections showing vascular system of the apple (McAlpine (1911a), Plate XXII. Courtesy of the Royal Botanic Gardens Library)

> About one month ago I received a letter from Professor Ewart offering some criticism of the two papers by you (McAlpine 1911a, b) on the grounds that you apparently thought that the existence of a vascular system in the apple pulp (McAlpine 1911a, Plates 21 and 22, see Figs. 11.1 and 11.2) was unknown to science, whereas N. Grew (in his *Anatomy of Plants*, 1682 AD – 230 years ago) had priority both in discovery and description.
>
> As the Council did not wish to be involved in barren controversy it was resolved that the Professor's letter should be received without further explanation, and he was so informed. A few days ago, I received another letter from the Professor enclosing a short paper setting out his criticism in detail and asking for publication thereof.

At this stage I was appiated [appealed] to as the Council naturally wished to hear 'both sides' and I was able to show that the 'vascular net' immediately beneath the skin of the apple and pear, and which has an important bearing on the origin of bitter pit, was not described in 1682 by Grew nor by anyone else.

The result of this was stated to me in a communication dated 18th September:

> I may add that the critical Note, mentioned in a previous letter has not been accepted.

2. Now as far as Professor Ewart is concerned, I would have allowed his letter received and his critical notes not accepted, to have remained in oblivion, but for a communication received from Dr Norris also referring to Grew's *Anatomy of Plants* (1682) and evidently inspired by Professor Ewart.

I may explain here that this ancient work by Grew is only known to a few botanists who delight in Ancient History and Professor Ewart himself writing in the Victorian Naturalist states; – 'Few copies of Grew's work are in existence. So far as I am aware no copy exists in Australia'. Judge therefore of my surprise when I received the following communication from Dr Norris dated 9th August:

Mr. D. McAlpine
440 Lonsdale Street, Melbourne.

> Averting to your interesting paper to the Linnaean Society, do you know anything of Grew's work, *The Anatomy of Plants* in which he deals with the subject with illustrations, though these are engravings as photography had not at the time been invented?
>
> The difficulty in the matter is that I have repeatedly heard fruit diseases discussed from the standpoint of attack through the circulation system and the whole basis of the discussion was that there was a fibre vascular network through the fruit pulp. Now I understand from your paper that the existence of this network was unsuspected up to the time of your discovery.

W. Perrin Norris

The bitter pit investigation is one requiring the expertise of the best brain-power that one possesses as it has been the despair of American, German, and British scientists and it is hardly fair that one should be harassed and annoyed by such trumpery criticism from the Chairman of the Advisory Committee.

3. I have now to refer to Dr Cherry and his reference to Grew at the meeting of the Advisory Committee on 30th October.

The Chairman invited me to be present towards the close of the Committee Meeting and while I was there Dr Cherry produced a copy of Grew's *Anatomy of Plants* from the University Library. He placed it on the table and triumphantly showed Plate LXVII, Fig. 4 as representing the vessels of the pear, of which I had claimed the discovery (McAlpine

1911b, Fig. 11.2). I was sorry for the ignorance displayed as the drawing only showed the well-known 'stone-cells' or 'grit' in the pear, and he would have found, if he had looked at the Explanation of the plate, that they were called by Grew in his ancient dialect 'The Testaceous Knots' [sclerid cells] (Fig. 11.3).

So that this gruesome discovery, initiated by Professor Ewart and followed by Drs Norris and Cherry, turned out to be a mare's nest after all!

I am sure there is no desire on the part of the Federal Government to hamper me unnecessarily in this investigation and I feel certain that when The Hon. Frank Tudor and yourself consider the matter on behalf of the States, these petty annoyances will cease in the future.

I have the Honour to be, Sir, Your obedient servant,

D. McAlpine

[PS] If you wish to see Grew's *Anatomy* at my home at any time, I can show you my private copy.'

A comparison of McAlpine's illustrations (Figs. 10.2, 10.3, 11.1, and 11.2) with those of Grew (Fig. 11.3) demonstrates that while McAlpine's was a serious study of the vascular system of pomes, Grew's inclusion of vascular strands was little more than notional, his main focus being on general anatomy and the presence of stone cells. While stone cells are still abundant in most quince varieties, their general presence in modern pears is generally much less than 70 years ago; the result of selection and breeding.

A month after the allegations of plagiarism were shown up for what they were, McAlpine (1912b, c) wrote to Minister Graham twice within a week drawing attention to Dr Norris's interference in the administration of the investigation. He told how he had been placed in an awkward position by having to provide monthly reports of his work to Dr Norris who was in communication with Professor Ewart. He noted from an announcement in the *Australasian* weekly review that Ewart claimed to be working on the same subject as McAlpine, but with out elaborating. He emphasised the point that he submitted a Progress Report once a year [as was required under his contract] and asked, 'Who would think of requiring a monthly report from an investigator say of Cancer or Consumption?'

McAlpine was keen to obtain the best available chemist to make comparative analyses of the chemistry of susceptible and immune varieties of apple. He had sought advice from Drs Norris and Cameron with the result that he had made arrangements to have the analyses carried out by a chemist in the Federal Chemical Laboratory. Then on 3rd December McAlpine received a letter from Norris telling him that Norris had written to each State Minister for Agriculture on the subject and that McAlpine was to keep the apples in cool store pending replies from the states. This angered McAlpine (1912c) who wrote again to Graham, drawing his attention to the fact that Advisory Committees and Advisory Councils were not given administrative functions under Federal Government procedures, and referred to the confusion it was causing.

Minister Graham responded to this in the context of McAlpine's concern over the printing and distribution of his First Progress Report, covered in the following section. A day later, Graham (1912) took up the matter with Dr Norris's Minister, The

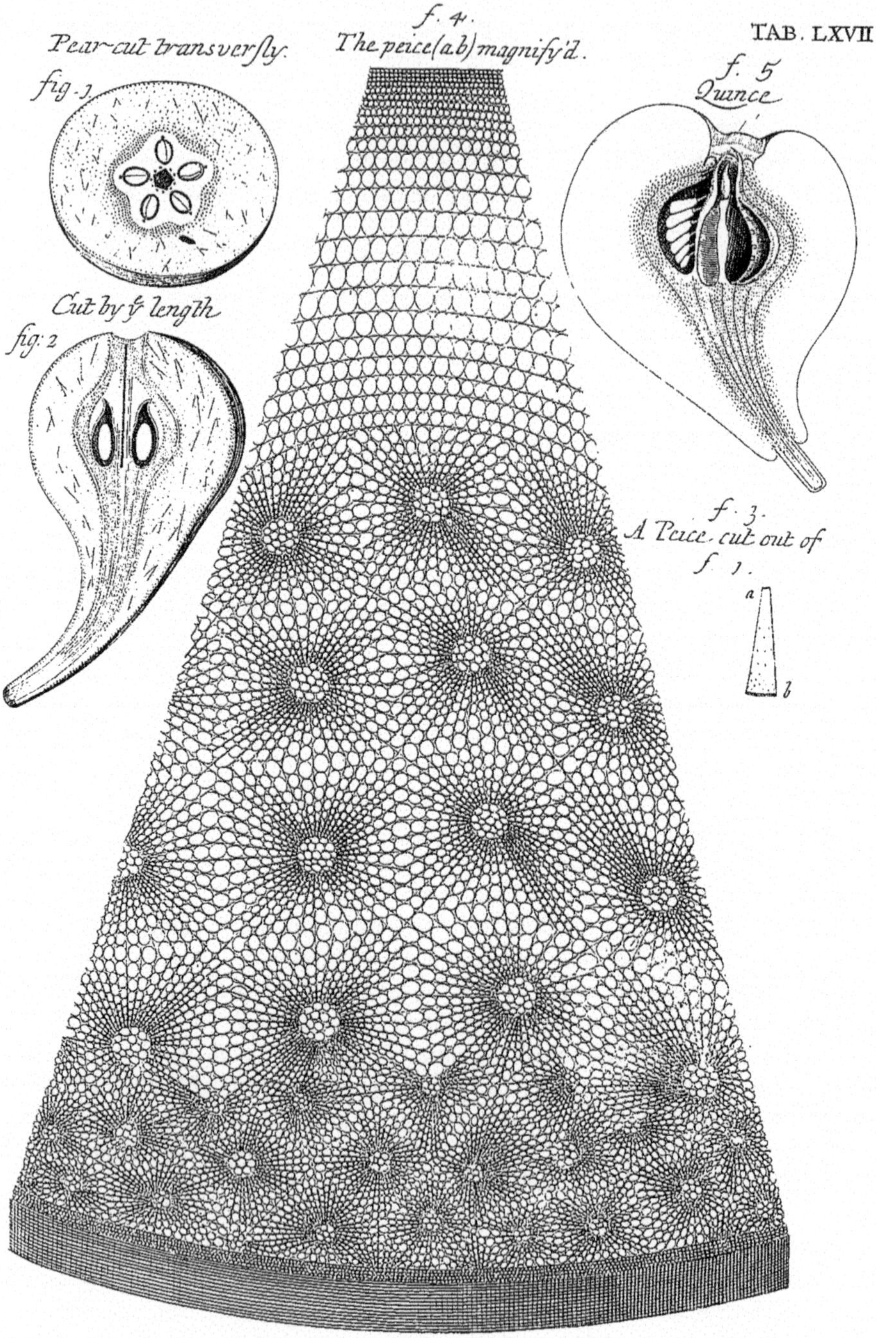

Fig. 11.3 Transverse and longitudinal sections of pear and quince. Main veins indicated, stippling highlighting aggregations of sclereide (Nathaniel Grew (1682) *Anatomy of Plants*, Plate LXVII, Fig. 4. Courtesy Special Collections, University of Melbourne Archives, Baillieu Library)

Hon. Frank Tudor MHR, Minister of Trade and Customs and, in a follow-up letter, assured Tudor that McAlpine was not wishing to override Dr Norris in any way. He suggested that Minister Tudor, Dr Norris and Dr Cameron meet with him to discuss problems with the printing of the First Progress Report and no doubt other problems including the selection of the chemist. On 12th December Minister Tudor (1912) responded to Minister Graham telling him that the muddle over the printing of the report was McAlpine's own fault as he had taken unauthorised action regarding its submission for printing. He closed the letter by more or less telling Minister Graham not to interfere in the Advisory Committee's realm of responsibility, and ended it with 'in connection with any future communications or arrangements connected with the investigation, Mr McAlpine shall address himself to Dr Norris who, in conjunction with Dr Cameron, is advising us of the subject in the intervals between the meetings of the Committee'. This last comment appears to refer to the requirement for monthly reports from McAlpine.

It is doubtful that the matter rested there. Mr Tudor had a reputation as a reasonable and just man, while Mr Graham and Dr Cameron were strong supporters of McAlpine. It is probable that there was a further meeting between Graham and Tudor at which the information from the NSW Linnaean Society and McAlpine's letters were tabled and discussed with Norris, who would have found it difficult to justify his actions. For whatever reason, Norris left for London soon after and signed no more correspondence from the Director of Quarantine's office in person, although some was signed in his name.

11.3 Distribution of the First Progress Report

By the beginning of 1913, McAlpine had received several enquiries about when copies of his First Progress Report would be available. He contacted the Government Printer who told him that 2,500 copies had been printed and bound and presumably delivered to the Department of Customs and Quarantine. On 17th February 1913, McAlpine wrote to Mr J.C. Hatton, an officer of the Victorian Department of Agriculture, warning him to ensure that he collected the full number of copies of the First Progress Report from the Department of Customs, and instructing him to dispatch 125 copies to orchardists who had supplied answers to his circular of questions (questionnaire!). He (McAlpine 1913) then wrote to Minister Graham, drawing attention to the delay in distributing the First Progress Report which he had sent to the Department of Customs on 1st July 1912, notwithstanding the statement by the Federal Treasurer in his Budget Speech on 3rd August, 'That the report be printed at once and made available.'

Daniel McAlpine and William Laidlaw were colleagues and friends, possibly having met in the lecture halls of Edinburgh in the early 1880s. After the Laidlaws arrive in Melbourne in 1908 the McAlpines offered their hospitality as they so often did (Fig. 11.4). It is therefore likely that McAlpine was aware that his progress

Fig. 11.4 The McAlpines enjoying Olinda's peace with the Laidlows in 1912 (From *Left* to *Right*, Isabella, Daniel and Erica McAlpine, William Laidlow, Veronica and Isabel McAlpine. Photo taken by Mrs Ludlow. Permission of the State Library of Victoria, Melbourne. State Manuscripts Collection MS 12772, Box 3555/3)

report should have been ready for distribution by the end of 1912 and was no doubt suspected the hold up was unwarranted.

The reason for the delay is not explained. Proof copies were available by late September because Norris (1912a) had sent one to William Laidlaw on 24th September 1912 when advising him that the first Advisory Committee meeting would be in October. He wrote again (Norris 1912b) on 17th October setting the meeting date for 29th and 30th of October, and advising Committee Members to bring their proof copy to the meeting, which he chaired. From 19th November 1912, all correspondence relating to the Advisory Committee sent from the Director of Customs' Office appears to have been handled by the Committee Secretary. He signed some letters in his own right, but mostly he applied a facsimile stamp of Norris's signature and initialed it. None of the surviving correspondence provides an explanation for this. Norris[A1] was reported as returning to London at the end of 1912. Mr Tudor (1912) wrote to Graham on 12th December telling him that 50 copies of McAlpine's report had been forwarded to the Minister of Agriculture in each of the six states and that he proposed to ask each how many more they required. Five hundred copies had been received at Customs in early December and the remainder was to be delivered in mid-March.

On 10th January, Dr Cameron sent a copy to Mr T.W. Kirk, Director of Orchards, New Zealand Departure of Agriculture, Industry and Commerce, who replied 3 days later (Kirk 1913), thanking him for it. It seems that McAlpine was not kept informed of any of this.

A letter was also sent from the Director of Quarantine's office on 20th March 1913 asking William Laidlaw, by then Government Biologist, to advise him how many additional copies he required, to which Laidlaw (1913) responded asking for 350 copies. Then in early May, Laidlaw requested the Victorian Orchard Supervisors to send him a list of the names of orchardists who were assisting McAlpine or had responded to his questionnaire. By the end of May the State's ten Orchard Supervisors had supplied him with 295 names, which on the 1st June, he sent to Dr Cumpston who had replaced Dr Norris in May.

Demands from the other states greatly exceeded the 50 copies made available to them and the Advisory Committee was unable to supply all the copies requested. It had grossly underestimated interest in McAlpine's investigation.

References

Graham, Hon. G (1912) Letter written [10-12-'12] to Minister Tudor proposing a meeting to consider problems surrounding the publication of McAlpine's First Progress Report and [unstated] problems between Norris, Cherry and McAlpine

Kirk TW (1913) Kirk, Director of Orchards, Department of Agriculture, Industry and Commerce, New Zealand wrote [21-1-'13] to Dr. Cameron thanking him for the copy of *The First Progress Report on Bitter Pit*

Laidlaw W (1913) Letter written [27-3-'13] to Dr. Norris requesting 350 copies of McAlpine's First Progress Report

McAlpine D (1911a) The fibro-vascular system of the apple (pome) and its functions. Proc Linn Soc NSW 36:613–625, 4 Plates, 21 Figures
McAlpine D (1911b) The fibro-vascular system of the pear (pome). Proc Linn Soc NSW 36:656–663, 3 Plates
McAlpine D (1912a) Letter written to Hon. Geo. Graham [c.11-'12] informing him of the spurious criticism from Dr. Norris and Cherry instigated by Professor Ewart
McAlpine D (1912b) Letter written to Hon. Geo. Graham [4-12-'12] objecting to Norris's requiring monthly reports on the grounds that it was not required in his contract and that the information contained was forwarded to Professor Ewart
McAlpine D (1912c) Letter written to Hon. Geo. Graham [9-12-'12] informing him that Dr. Norris was acting outside his brief by interfering in administrative arrangements by changing arrangements already made by McAlpine for chemical analyses to be made at the Federal Chemistry Laboratories
McAlpine D (1912d) The fibro-vascular system of the quince compared with that of the apple and pear (Pome). Proc Linn Soc NSW 37:689–697
McAlpine D (1913) Letter written [4-3-'13] to Hon. G. Graham MLA drawing attention to the delay in distributing the *First Progress Report* which he sent to the Department of Customs on 1st July 1912, not withstanding the Statement by the Federal Treasurer in his Budget Speech on 3rd August, "That the report be printed at once and made available." He had checked with the Government Printer who had told him all 2,500 copies had been printed and bound
Norris WP (1912a) Letter [24-9-12] written to W. Laidlaw, and other members of the Advisory Committee, with copies of the First Progress Report and foreshadowing a date for the first Advisory Committee meeting
Norris WP (1912b) Letter [17-10-12] written to members of the Advisory Committee advising date time, and venue of the first Advisory Committee meeting
Tudor, Hon. FG (1912) Letter written to Hon. G. Graham [12-12-'12] blaming McAlpine for the confusion over the printing of his First Progress Report and backing Dr. Norris authority to handle administration of the investigation

Chapter 12
The Second Year 1912–1913

12.1 The Second Progress Report

In his second report McAlpine (1913) focused on factors which he believed contributed to bitter pit, as well as the issue of most concern to the public. In regard to the latter, he enlisted the help of Dr Rothera[A1] and his student R.H. Greenwood, biochemists at the University of Melbourne, to test the poison theory still supported by Professor Ewart. McAlpine supplied the biochemists with batches of pit-affected and healthy apples for chemical comparison. They found no chemical differences in starch grains from pitted and non-pitted tissue, no poisonous metals in tissue from pitted apples even in trace amounts, and no differences in the cell-wall chemistry of tissue from pitted and non-pitted apples. He also focused on looking for differences between the bitter-pit-susceptible *Cleopatra* and resistant *Yates* varieties (Fig. 12.1).

McAlpine noted that he did not subscribe to Sir John Gilbert's[A1] view that plant growth was essentially the chemical offspring of the soil. McAlpine believed each plant had its own constitution. He implied that this determined each variety of apple's capacity for nutrient uptake. He gave a lengthy explanation, as far as it was understood, of the role of osmosis in the uptake and transport of water in plants and stated that during the growing season the average apple tree took up and transpired about 250 gal [1,200 l] of water each day. He also named 10 experimental sites; Kelso near Bathurst in New South Wales, Bacchus Marsh, Burnley Horticultural Gardens, Deepdene, Diamond Creek, Harcourt and Leongatha in Victoria, Rokeby and Scamander in Tasmania, and Blackwood in South Australia, most within a 2 h horse and buggy drive of a railway station. Each was run under his direction by local orchardists and agricultural officers, while he visited each site as often as possible.

From data gathered from the many experiments, he noted that orchard management that ensured high levels of pollination lowered the risk of small crops of large fruit and helped contain the incidence of pit, that there was no evidence that bites from harlequin bugs, thrips or mites caused bitter pit, and bacteria and fungi

D.G. Parbery, *Daniel McAlpine and The Bitter Pit*,
DOI 10.1007/978-3-319-09552-3_12

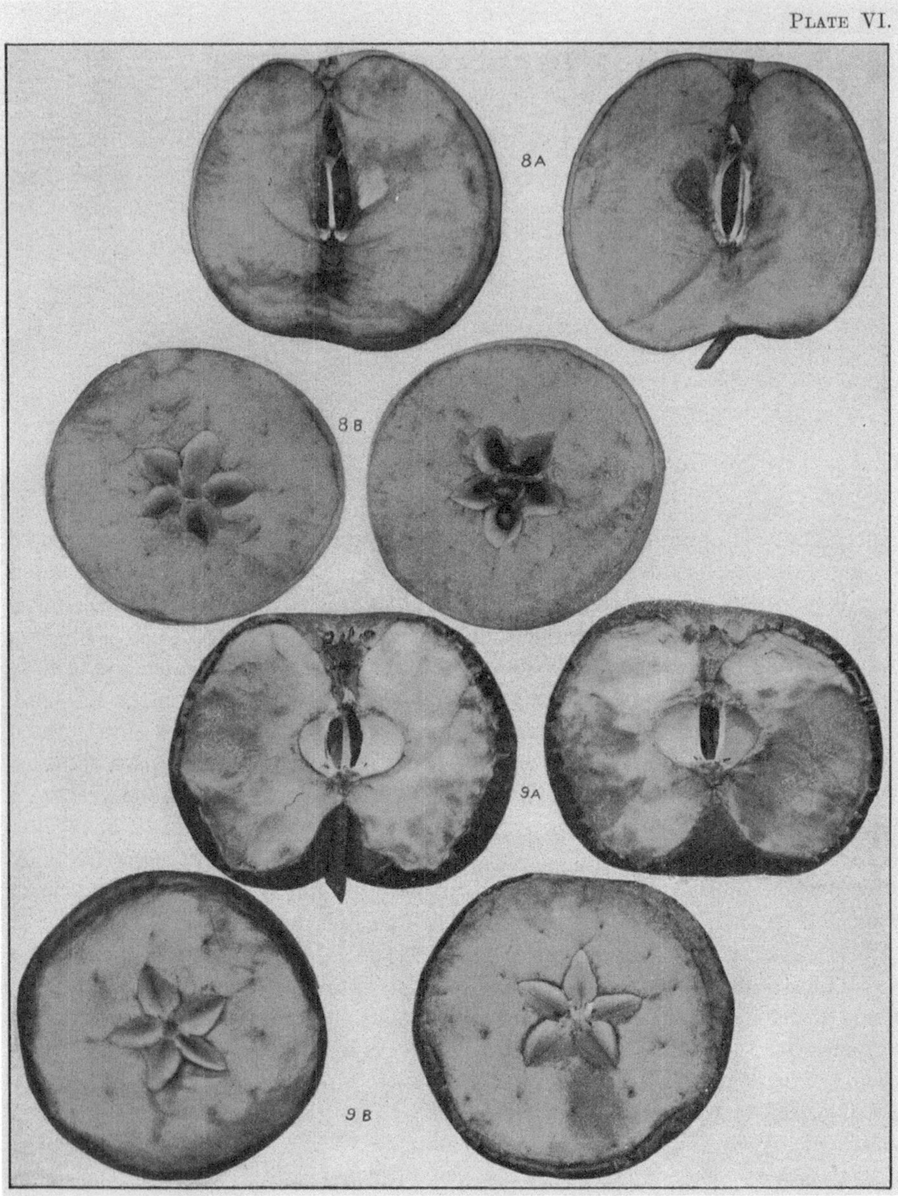

Fig. 12.1 Median TS and LS sections of *Cleopatra* (*upper set*) and *Yates* (*lower set*) apples, dried to highlight the arrangement of the primary vascular bundles (McAlpine's Second Progress Report Plate VI (1913). Courtesy of the Royal Botanic Gardens Library)

were not involved. His findings also supported his earlier view that there were considerable differences between apple varieties in their susceptibility to bitter pit, and that heavy pruning resulted in fewer larger more seriously pitted fruit than light pruning. He reported that in most cases pitting appeared first at the 'eye' (calyx) end of fruit and ascribed this to the greater number of pores in the skin in that region and therefore the greater evaporation rate of water. He described the vascular network that developed in fruit and noted that the first appearance of pit was just under the skin at vein endings. He referred to a paper written by his friend Professor Whetzel (1912) who had also concluded that the condition was associated with the abnormal distribution of water in fruit.

12.2 Monthly Reports

On 4th December 1912, McAlpine (1912a) wrote to Minister Graham objecting to being required by Dr Norris to provide monthly progress reports. By then he had realised that Drs Norris and Cherry were under the influence of Professor Ewart. This was of serious concern because he knew that Ewart was attempting to undermine his investigation. December was the fourth month into the second year of the investigation, and McAlpine had sent the chairman a single foolscap page report of the month's activities, which he continued to do until 1916. A copy was sent each month to Dr Cameron, and possibly to the other Directors of Agriculture.

On reading these reports, without having read the above letter, one is struck and somewhat frustrated by the absence of any detailed results. The reports merely tell what he did, but not what he discovered. Having read the letter (McAlpine 1912a) it became clear that McAlpine was depriving Norris of opportunities to feed Ewart preliminary results, which he knew were manipulated in ways that undermined his own conclusions in the eyes of some members of the Advisory Committee and of his political masters. In spite of the lack of outcomes in these reports, abstracts are provided here to demonstrate the amount of personal effort McAlpine put into the investigation in addition to setting up experiments, analysing results, and keeping the Advisory Committee fully informed. Each was typed, addressed to the Advisory Committee, and began in the formal manner of the time: – 'Gentlemen, I have the honour to inform you that during the past month (named) I have been principally engaged in the following:'. Each report related to the previous month's activities.

December 4th 1912 In order to test the amount of water given off by apples in the course of a season, he enclosed fruit still attached to the tree, in a glass bulb. *Autumn Pearmaine* and *James Grieve* were the varieties chosen; the water was run off weekly by means of a tap, and measured. *Annie Elizabeth* was cinctured for the last time on 6th November just as the petals were falling, and another tree was compressed with fencing wire. The pruning experiment on *Cleopatra* at Deepdene was visited and the condition of the fruit under each treatment noted, viz., – severe, light, leader, and no pruning. He visited the manurial experiment at Box Hill, noting the nature of the setting [relative abundance of new fruit] and the stage of maturity of the fruit in each plot. At the Doncaster Cool Stores the varieties *Esopus Spitzenberg* and *Cleopatra,* placed there on 10th and 25th April respectively, were free of bitter pit and fresh in flavour. It was arranged to have the same varieties tested next season and shipped to London on the *Argyleshire* and the *Osterley*. The earliest appearance of bitter pit was observed on the apple variety *Borrowdale* on 29th November. Proofs of the First Progress Report were corrected and the correct printing of the 33 plates overlooked (meaning 'looked over' not 'missed'?). Each monthly report ended in the same way.

January 4th 1913 During the last month various orchards were examined for the earliest appearance of bitter pit. Both *Pomme de Neige* and *Lady Carrington* were sent from New South Wales as among the earliest, but the latter variety turned out to be *Thorle Pippin,* which was also badly pitted in the Burnley Gardens. Experiments there were carefully inspected and records of progress taken. McAlpine inspected *Cleopatra* and *Yates* apples which were growing alongside each other in Hatfield's orchard, and which would be used for comparative chemical analysis [of bitter-pit-susceptible and immune varieties respectively]. A pressing machine was obtained for extracting apple juice for analysis, with the weekly measurement of water expired by growing apples to continue until the end of season, and a circular was submitted to horticultural officers containing a list of observations requested from them. The most important feature of the month was a visit to Tasmania where the principle orchards of the Huon and Derwent Valleys were inspected in the company of the Director and Horticultural Expert, comparing the relative susceptibility of the different commercial varieties of apples to bitter pit.

It is of interest to note here that on 8th January 1913, McAlpine received a letter from Sir Oliver Lodge[A1] (1913), a fellow student of physics at the Royal School of Mines in 1874, and a pioneer in the discovery of radio waves and the development of X-ray. McAlpine was already exploring the possibility of using X-ray to determine the presence of incipient bitter pit in apparently healthy fruit and had evidently written to Lodge for his views and advice.

February 3rd 1913 *Cleopatra* and *Yates* apples were examined at Box Hill and Burnley, and samples were supplied to Wilkinson for analysis. The thickness of the skin determined by microscopic examination, the number of stomata and other features were compared. He visited Mr Cowan's irrigated orchard at Bacchus Marsh where he arranged to have one half row of a variety irrigated and the other not. An important experiment was under way at the Australian National Apple (ANA) Exhibition where the Government had erected a miniature cool store, which would be operated at fluctuating temperature between 32 and 40 °F [0–4.5 °C] for the duration of the exhibition, while a case of apples from the same tree was kept at a constant temperature of 32 °F in a cool store at Doncaster. Arrangements had been made with Dr Rothera to compare the chemistry of pitted and non-pitted apples supplied to him. During the coming month McAlpine intended dispatching a consignment of apples to London, in a test of keeping quality.

The fate of this consignment is dealt in Sect. 12.4, under the heading 'The Voyage of the *SS Somerset*'.

March 4th 1913 McAlpine was occupied examining apples from the ANA miniature cool room and from Doncaster, and supervising experiments at Box Hill, Deepdene and Burnley. He obtained 15 cases of *Cleopatra* apples from the Harcourt orchard of which 14 cases were dispatched to the Agent-General in London aboard the *SS Somerset* [which was equipped with cool storage facilities], in order to determine the development of bitter pit and the keeping quality of the apples after removal from the steamer. The remaining case was kept as a control in the Doncaster cool store. The miniature cool store at the ANA Exhibition afforded a splendid opportunity for testing the effect of fluctuating temperature on the development of pit. The amount of pit that developed in the 26 boxes of apples was carefully determined in his office. In one instance, all 138 apples in a case of *William's Favourite* were affected. In the pruning experiment at Deepdene, the highest incidence of bitter pit occurred under severe pruning and the least under no pruning. On a visit to Blackwood in South Australia he inspected experimental plots and toured surrounding orchards. Microscopic sections of the skin of *Cleopatra* and *Yate*s apples, as well as other characteristics, were being examined. Early in March he foreshadowed a visit to Western Australia to examine experimental plots.

April 5th 1913 He visited the West where a number of experiments were being conducted. The proprietors of the Mount Barker Estate afforded him every assistance, ensuring the results would be of the highest importance. All the apples were carefully weighed, the windfalls were likewise allowed for. A number of the orchards visited in the district were fully 200 acres in extent, and since the favourite apple was *Cleopatra*, the orchardists were particularly interested in the bitter pit experiments. At Deepdene the apples were also gathered from the experimental plots and the results showed that the pit developed least in the unpruned plots. At Burnley Gardens the trees that were cinctured and compressed with wire were picked and the proportions of pitted apples noted. Apples in the 26 lots taken from the Doncaster Cool Store showed no external signs of pitting. Apples in the miniature cool store at the ANA exhibition, put there to display the effects of fluctuating temperature on stored apples, were removed to determine the proportions of pitted and clean fruit. At Mr Thiele's Doncaster orchard there were two trees, each with a number of varieties grafted on the one stock. The apples borne by each variety were carefully counted, and while some [varieties] were badly pitted, others were perfectly clean.

May 3rd 1913 During April, McAlpine was busy recording results from numerous experiments. He visited Campbell's Creek to examine grafts from pear trees. At Burnley, apples were picked from trees variously pruned and the pitted separated from the clean. The proportions of pitted and clean *Sturmer* apples picked from the irrigation experiment at Bacchus Marsh were determined. Also the proportions of clean and pitted apples from the manurial experiment at Mr Hatfield's orchard, Box Hill, were recorded. He visited Douglas's orchard at Harcourt where *Cleopatra* apples were comparatively free from pit on trees that had been well manured, frequently watered and unpruned. A case of *Cleopatra* apples kept at the Doncaster cool store was opened on 10th April, the date of arrival of the *Somerset* in London carrying the 14 cases for the Agent-General. The fruit was commercially free from pit.

June 5th 1913 During May he was largely engaged in tabulating the results of experiments in the different States and making deductions from them. At Burnley Gardens a number of different varieties of apples were kept in store and he had been examining them periodically to track the development of bitter pit. It was also noted that the cinctured limbs of apple and the wired limbs of pear trees dropped their leaves sooner than non-constricted limbs. He visited Dunolly where bitter pit was very bad on *Cleopatra* apples and found some treated trees that were free of it. He also visited Thiele's orchard at Doncaster where quinces were badly pitted, but they were also badly damaged by a severe hailstorm. At Douglas's orchard at Harcourt some of the trees with small crops were badly pitted, while others with large crops were comparatively free. Photos were taken to show the difference. He foreshadowed visiting New South Wales in June in order to inspect various orchards in the Murrumbidgee irrigation area. The idea was prevalent that fruit from non-irrigated trees was less pitted than that from irrigated trees.

July 5th 1913 During June, McAlpine visited New South Wales and portions of Victoria in connection with experimental work. He obtained the results from the experimental orchard at Bathurst from the various plots in manuring, pruning and irrigation, and arranged the work for the coming season. At Harcourt various interesting apple trees were examined and photographed – an *Annie Elizabeth* tree of which a limb had been cinctured, a *Bramley,* not pruned for the last 20 years, and yet the previous year yielding 26 bushels [1,092 lbs = 496 kg] of fruit, and a casual seedling named *Sunbeam*, which had a little pit, although it was asserted by some that seedlings on their own roots were free [of bitter pit]. Further evidence of there being no connection between spraying and bitter pit was supplied by the Bureau of Microbiology, New South Wales. The report for the year was being prepared and manures for the experimental plots in each state were being dispatched.

12.3 Long Distance Transport of Meat, Fruit and Vegetables

The background history of the need for low temperature facilities for long distance transport of meat and fresh produce is relevant to McAlpine's investigation which paved the way for enabling Australian apples to be supplied to the London market in marketable condition. For two decades after the first Australian apples were exported to London in 1891, the chance of having consignments of popular eating apples such as *Cleopatra* arrive in wholesome condition was low because of the development of bitter pit during the voyage. By 1912, however, McAlpine had confirmed the benefits of low temperature storage, and was ready to investigate the benefit of using cool storage during overseas transportation, which by then was fairly reliable.

The first known cool stored freight over long distances was in the USA where in 1857 a consignment of beef was packed into railway vans lined with blocks of ice and taken from Chicago to the east. Ten years later a form of refrigeration was trialled with mixed success, then in 1869 salted ice was trialled with such success that chilled beef could be carried much further than previously, and by 1877 this practice enabled consignments of beef to be carried from the USA to Britain. The level of success of these shipments depended on longitude, distance and weather, but in spite of some losses the initiative brought great commercial success to the pioneers. Interest in cool storage of fruit was not new, in fact it received attention at the 1909 Perth conference, but it had not be tested for it effectiveness during long distance transport.

The use of blocks of ice to preserve food had been used since pre-Roman times, but had obvious limitations, especially for long distance transport through tropical climes, even with the addition of salt to the ice. Interest in the development of refrigeration emerged in 1756 at the University of Glasgow when William Cullen used a pump to create a partial vacuum over a container of diethyl ether which then boiled, absorbing heat from the surrounding air. He succeeded in producing a small amount of ice. Others toyed with this process until 1805 when an American inventor, Oliver Evans, designed but never built a refrigeration system based on vapour-compression. Several scientists developed systems based on this principle although none was developed commercially for over 40 years.

In 1851, James Harrison, an Australian living in Geelong, produced an ice-making machine, and in 1856 took out a patent for it, an ether liquid-vapour compression refrigerator. By 1861 this had been adopted widely by breweries and meat packing houses and paved the way for the development of refrigerated cool storage and transport, which, during the 1870s encouraged Australian, Argentine and American entrepreneurs to experiment further with long distance transport of meat. Initially, success was not assured. Failure in the logistics of getting frozen meat into ships' cool store from land-based cool stores before it thawed sometimes destroyed consignments. Breakdowns in the refrigeration machinery, especially at sea, were a serious problem. The first remarkable commercial success came when William Davidson installed a compression refrigeration system in the sailing ship *Dunedin*, and in 1882 sent a cargo of beef, mutton, lamb and other meats from New Zealand to London, landing it there in perfect condition after a voyage of 98 days. The quality of New Zealand mutton and lamb was immediately accepted as better than Australian meat, a benefit for New Zealand that lasted over 100 years. The success of the voyage of the *Dunedin*, however, opened the way for a boom in meat and dairy products from Australasia and Argentina as well and opened the way for the development of fruit exports.

Problems involved in transporting chilled fruit and vegetables took longer to overcome than they did for meat. A fundamental reason may have been the need to develop markets in Europe for fruit and vegetables from the antipodes. It may also have taken longer to develop the necessary protocols for cooling different fruits and vegetables. Early experiences would have sounded notes of caution. In 1866 a successful consignment of chilled strawberries was sent east from Chicago in ice-cooled vans, with encouraging results. However, during the 1870s, when the first refrigerated rail wagons were introduced, the results of transporting certain vegetables were disastrous. For example, the first load of lettuce transported over a long distance in the USA arrived at its destination as an unsaleable thawed out mess of mush, having been frozen. By the late 1880s, however, oranges and pome fruits were being successfully transported to markets by rail from Florida and California in refrigerated vans, and by 1901 such vans also provided controlled CO_2 facilities for the transport of bananas. By 1904, G.H. Powell, also in America, had devised a system for pre-cooling pome fruit before it was transported, but apparently not in cool storage transport.

12.4 The Voyage of *SS Somerset*

In the 1870s and 1880s, Tasmania was producing large volumes of apples, most of which were shipped to other Australian colonies. By the late 1890s the bulk of Australian apples exported to the UK was from Tasmania, soon best known as the *Apple Isle*. Increases in overseas exports were driven by three factors: the increasing use of steam ships, the availability of refrigerated shipping, and better prices for apples on the London market than at home, especially during the northern hemisphere winter. By 1905, South Australia was exporting 150,000 bushels [2,813 t] of apples to London annually. By that time, however, the most popular varieties such as *Cleopatra*, which commanded the highest prices, often arrived in unmarketable condition because of the development of bitter pit during the voyage. The problem was not confined to the Australian states, but was a general problem experience by all countries exporting pome fruit over long distances. Nor was it confined to one variety. *Baldwin* apples in the USA and *Jonathans* in Australia were also problematic as were many other varieties. Such varieties appeared sound when fruit for export was selected, packed and loaded on to ships, but in affected fruit the symptomless incipient bitter pit developed in about 1 month under ambient conditions.

During the decade prior to the bitter pit investigation, McAlpine had become aware of this aspect of the bitter pit problem, so that while he was studying the benefits of putting freshly gathered fruit immediately into cool storage, he also turned his mind to the prospect of reducing the deterioration of fruit during the 6 weeks or longer transit time to London. By 1913, within the British commercial fleet there were 230 vessels with a total refrigerated capacity of 440,000 t. Thus, such transport was readily available, and on 17th February 14 cases of *Cleopatra* apples were loaded on the *SS Somerset* at Williamstown Pier. The ship sailed on the 18th February and docked in London on 10th April, after a voyage of 7 weeks and 2 days. Arrangements had been made for the Agent-General in London, Sir Newton James Moore[A1], to have the apples inspected on arrival, then again after 2 weeks stored under ambient conditions, and again after 1 month under ambient conditions. Following each of the three inspections when every apple had been examined, the Agent-General forwarded the results to McAlpine (Brisley 1913). McAlpine compared the incidence of bitter pit in the transported apples after the 51-day voyage with samples of the same consignment kept in the Doncaster cool store in Australia over the same period. The results of the *SS Somerset* consignment are given in Table 12.1.

In an earlier Chap. 7 (Sect. 7.3), it was stated that the average weight of four small apples was one pound, equivalent to three medium sized apples, or two large

Table 12.1 Condition of apples in the *SS Somerset* cool store consignment, on arrival in London [10th April], then after 2 weeks [25th April] and 4 weeks [8th May] held at ambient temperature

Condition	18th Feb	10th April	25th April	8th May
Sound	1,900	1,719	1,657	1,592
Pitted	0	178	239	296
Rotten	0	3	4	12

apples. It seems that McAlpine chose extra small apples weighing closer to five per pound. Assuming that the 138 very small apples in the box retained by McAlpine was typical of the consignment sent on *SS Somerset,* the consignment would have been made up of approximately 1,900 apples. On this basis, after 7 weeks in transit, 89.5 % of apples were marketable. Two weeks later, during which the apples were held at ambient temperature during London's early spring, their marketability was 87.2 %, and after 4 weeks at ambient spring conditions, 83.8 % were still marketable.

McAlpine's advice was to market fruit within 2 weeks of its removal from cool store. Six years later, however, he extended his advice for preserving the quality of apples during long distance transport (Chap. 17, Sect. 17.2).

12.5 The Advisory Committee's Second Report

McAlpine (1913) submitted his second progress report in October 1913, again liberally illustrated with photographs and a fine watercolor by Brittle bank (Fig. 12.2). The second meeting of the Advisory Committee was set for the 21st and 22nd January 1914 (Crathers 1914a). W. Craters, Secretary to the Advisory Committee, had advised members of the date and reminded them to take their proof copies of the Second Progress Report with them (Crathers 1913). This meeting was to have been held initially in November 1913.

Following the meeting, the Advisory Committee (Crathers 1914b) put nine recommendations to the Minister for Customs and the Minister for Agriculture as follows:

1. That the comparison between susceptible and resistant varieties of apples should be extended to include more than one variety of each and they recommended comparing three susceptible varieties, *Cleopatra, Northern Spy* and *Annie Elizabeth* with three resistant ones, *Yates, Stone Pippin* and *Dunns' Seedling*. Furthermore, they suggested that in order to examine the influence of climate on relative susceptibilities to bitter pit, comparative analyses should be conducted in the Government laboratories in each state.
2. They were not convinced by McAlpine's claim that the cause of bitter bit involved interference with the development of the vascular system and the collapse of cells, and recommended that further evidence be supplied to support McAlpine's hypothesis.

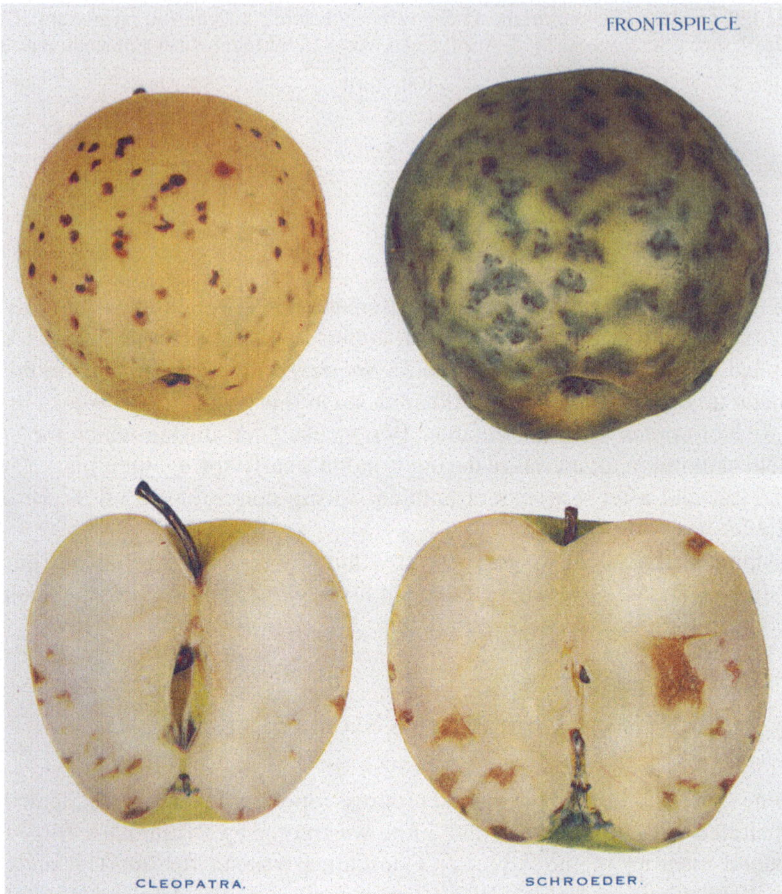

Fig. 12.2 Symptoms of bitter pit in *Cleopatra* (*left*) and *Schroeder* (*right*) apples (McAlpine's Second Progress Report (1913), Frontispiece. Courtesy of the Royal Botanic Gardens Library)

3. The Committee repeated its previous recommendation that going forward, experiments should include as many trees as practicable in order to substantiate the conclusions drawn in the Second Report.
4. The Committee considered that in view of the importance McAlpine attached to the rate of transpiration in affecting the incidence of bitter pit, that further experiments were needed with a view to controlling it in orchards. They suggested McAlpine should include experiments in root pruning, the effect of saw cuts into branches and the sheltering of susceptible trees from sun and wind.
5. They suggested that X-ray examination should be tried to determine if it were possible to detect bitter pit developing prior to apples being stored with the object of ascertaining whether the disease actually developed *de novo* after cold storage.

6. They were not satisfied with the standard of the irrigation experiments and reaffirmed the contents of Recommendation 5 of their first report.
7. The Committee directed this recommendation specifically to State Authorities. That in order to help Mr McAlpine evaluate the significance of shelter on the incidence of bitter pit, State Authorities be asked to allow field officers to select sheltered orchards for special observation and to inform Mr McAlpine of such places.
8. This recommendation urged State Authorities to ensure that the benefits of McAlpine's findings pertaining to the voyage of *SS Somerset* were put into effect. The committee made three recommendations regarding cool storage during transportation; **(i)** that each exporting State instigate as regular practice, the placing of a sample case from each export consignment in cool storage here to be compared with the condition of the shipment on its arrival in London. **(ii)** that the samples stored and conveyed be examined and compared for the prevalence of bitter pit. **(iii)** that the samples be kept for 1 month after coming out of cool storage, under conditions comparable to those in the hands of retailers, and they be again examined and compared for the presence of bitter pit.
9. That experiments be continued as to the regularity of fruiting so that heavy crops in one season and light crops the next might be avoided.

While the suggestions and recommendations of this and their first report appear reasonable, they were made when the third season was more than half way through and when, in terms of the original contract, there was only one more season to run. Even so, some were already being implemented prior to the January meeting and most of them were conducted over the following 2 years.

McAlpine's comments made in a letter to Dr Cameron a year later (McAlpine 1914) indicate that the analyses of the juice from the various varieties nominated by the Advisory Committee was done by the chemists in each state as arranged by Norris. Strict procedures were laid down to ensure that the analyses were as uniform as possible in terms of which fruit was picked and treated in the laboratory. Some of his concern over uniformity of analyses may have been avoided if Norris had not changed McAlpine's initial arrangement with the Federal Chemistry Laboratories (McAlpine 1912b). He also explained to Dr Cameron that the varieties selected by the Advisory Committee were not entirely satisfactory since the variety *Stone Pippin*, listed as resistant to pit by the Advisory Committee, proved susceptible when grown in some locations, whereas varieties such as *Cleopatra* and *Yates* behaved consistently (McAlpine 1914).

References

Brisley AJ (1913) Letter [May 1913] regarding condition of apple in the consignment on *SS Somerset*

Crathers W (1913) Letter written to W. Laidlaw [18-10-13], from the Secretary of the Advisory Committee regarding the printing of the second progress report and foreshadowing an Advisory Committee meeting for mid to late November 1913

Crathers W (1914a) Letter written to W. Laidlaw [13-1-'14], notifying him of a meeting of the Advisory Committee meeting at 10:30 am on Wednesday 21st January at the Quarantine Bureau, 51 Spring Street Melbourne
Crathers W (1914b) Letter written to W. Laidlaw [29-1-'14], with a draft Progress Report and concerning a baseless allegation that Burnley Gardens were established on the site of an old rubbish-tip
Lodge O (1913) A letter [8-1-'13] sent by Sir Oliver Lodge to McAlpine regarding the possibility of detecting incipient bitter pit in apples by x-ray
McAlpine D (1912a) Letter written to Hon. Geo. Graham [4-12-'12] objecting to Norris's requiring monthly reports of his progress on the grounds such a requirement was not included in his
McAlpine D (1912b) Letter written to Hon. Geo. Graham [9-12-'12] informing him that Dr. Norris was acting outside his brief by interfering in administrative arrangements and cutting across arrangements already made by McAlpine for chemical analyses to be made at the Federal Chemistry Laboratories
McAlpine D (1913) Bitter pit investigations. Second progress report – the cause of bitter pit: its contributing factors, together with an investigation of susceptibility and immunity in apple varieties. Melbourne, Government Printer 224 pp, bibliography 96 pp, and 61 plates & maps
McAlpine D (1914) Letter to Dr. Cameron [29-9-'14] reporting on the analysis of juice from reputedly susceptible and resistant apple varieties
Whetzel HH (1912) Baldwin Spot or Stippen. In: Proceedings of the New York State Fruit Growers Association 11:28:34

Chapter 13
The Third Year 1913–1914

13.1 Monthly Reports

Among the documents found, only those recorded here have been located even though the covering letters sent with many of them were kept. Even so, the reports of his activities between late 1913 and the end of 1914 provide some idea of the effort he put into the investigation.

August 4th 1913 During June, McAlpine set up experiments for the season ahead. He forwarded specific manure/fertiliser mixtures to each state for various experiments. Pruning experiments were conducted at Burnley Gardens and Deepdene, where blocks of trees were either severely pruned, lightly pruned, leader pruned or not pruned. In connection with the testing of rootstocks, trees from Cole's Nursery were planted, with *Annie Elizabeth*, *Cleopatra* and *Jonathon* grafted upon brabs and crabs worked upon *Northern Spy* stocks [I assume 'brabs' and 'crabs' were shorthand for *Bramley* and *Crab Apple* scions already grafted on to *Northern Spy* root stocks, so as to separate the final graft from direct contact with the root stock, DGP]. The analyses of the juices of *Cleopatra* and *Yates* apples were received from Mr Wilkinson and a report written. In an attempt to identify incipient bitter pit in apparently healthy fruits of *Cleopatra* and *Yates*, x-ray photographs were taken of clean and pitted fruit, which would appear in his next report. The Second Progress Report was being written.

September 4th 1913 During the last month, McAlpine was engaged in establishing experimental plots in various states. In New South Wales, not only were pruning experiments carried out at the Bathurst Government Farm, but also on *Cleopatra* trees at Mr Edgell's private orchard. In Tasmania he had access to *Cleopatra* trees in North and Brady's orchard on the Tamar, where pruning experiments were also begun. The Director of Agriculture in Hobart advised him of an orchard at New Norfolk with over 1,000 *Cleopatra* trees, where sometimes up to 80 % of fruit was pitted. He visited there the following week to carry out manurial and pruning

D.G. Parbery, *Daniel McAlpine and The Bitter Pit*,
DOI 10.1007/978-3-319-09552-3_13

experiments, and to cincture with a zinc bands trees that yielded poorly. He had also visited Mr Smith's orchard at Deepdene, Mr Hatfield's at Box Hill, and Burnley Gardens where he applied zinc bands to trees not bearing well.

October 4th 1913 During the past month he completed and forwarded the Second Progress Report, entitled 'The Cause of Bitter Pit – its Contributing Factors together with an Investigation of Susceptibility and Immunity in Apple Varieties', to the Minister for Customs (McAlpine 1913). He visited an orchard at New Norfolk, Tasmania, where fruit of *Cleopatra* were badly affected with bitter pit, and carried out experiments in pruning and the use of zinc bands. Apparatus was being fitted up at Burnley Gardens to measure the amount of water being transpired by growing apples so as to have the experiment as close as possible to natural conditions. Pips from the cross between *Cleopatra* and *Yates* had been planted, and for comparison the parent varieties were being grown alongside them. The various experiments under way were being closely observed, and the time of blooming carefully noted. The practical issue at the time was the prevention of pit in growing apples, since post-harvest pit development was completely controlled in store by keeping the fruit at a sufficiently low and uniform temperature. Numerous experiments with different stocks were beginning to show results, and the cincture, constricted and broken branches were being studied in relation to the rest of the tree. The Second Progress Report was sent to the Government Printer about mid October (Crathers 1913).

November 3rd 1913 During October, he arranged experiments for the coming season, mostly at the Burnley Horticultural Gardens, and examined specimens of apples from cool store with skin diseases. The blooming of the different varieties had been observed, with particular attention paid to those that had been 'rung', constricted, or girdled with the zinc band. Trees in full bloom were pruned to compare the result with that of winter pruning in trees. The numerous rootstock experiments were examined. Since transpiration appeared to play such an important part in the onset of bitter pit, he was conducting an important experiment, in conjunction with Mr Scott, Agricultural Chemist, on the transpiration of apples at different stages of growth. Once firmly set, young apples were enclosed individually in glass bulbs fitted with calcium chloride tubes, so that the amount of moisture given off at various stages could be determined. He had also visited South Australia and, with Mr Quinn, visited the government orchard at Blackwood. The experiments being carried out there were more numerous, varied and extensive than in any other Australian State.

December 1st 1913 During November the experimental work in connection with bitter pit had been specially attended to. The transpiration of the growing apples enclosed in glass flasks was working satisfactorily, and Mr Hunt, Federal Government meteorologist has kindly supplied a maximum and minimum thermometer, in order make the records as complete as possible. The first appearance of pit has been comparatively early in the season, owing to the extremely variable weather, and it was found on the following varieties on 18th November – *Williams*, *Red Must*, *Red Normanby* and *Forest Styre*, the apples being about the size of filberts [hazel nuts]. The 54 varieties of apples on various stocks, enclosed in screens, had been sprayed, after an examination for bitter pit. The disease was found on the apples produced by *Cleopatra* and *Duke of Clarence*, each grafted onto *Cole's Paradise* stock. A *Northern Spy* was chosen as a stock and grafted in September with four pit-liable varieties; *Cleopatra, Annie Elizabeth, Cox's Orange Pippin* and *Bismark.* There were two grafts of each and they took in every case. Each variety had already fruited excepting *Cleopatra.* He was leaving in December to inspect the experimental plots at Stanthorpe in Queensland, the Experimental Farm at Bathurst, and Yanco in the New South Wales Murrumidgee irrigation area.

January 3rd 1914 McAlpine's visits planned for Queensland and New South Wales were made and the various experiments were inspected. In spite of no rain for seven weeks prior to his visit to Stanthorpe, there was sufficient moisture in the soil to tide them over. He also inspected the experiments at Mr Hatfield's orchard at Box Hill. Arrangements were made for a supply of half-grown *Cleopatra* and *Yates* to be analysed and compared with similar varieties from an irrigation area. A record is being made of the numerous varieties of apples at Burnley Gardens and the respective rootstocks they were on in order to determine the relation between the nature of the stock and the amount of pit. The various experiments were being monitored and windfalls carefully recorded.

February 4th 1914 An inspection was made of the condition of fruit at Bacchus Marsh where there was a row of *Sturmer Pippins* in which one half was only irrigated once and the other several times. McAlpine visited Toomuc Valley to inspect one of the largest apple orchards at North Pakenham in Victoria where there were trees of *Champion* apples very liable to bitter pit. These had been left unpruned and the development of the pit monitored. He was looking for any striking chemical differences between irrigated and non-irrigated apples grown under similar conditions. He obtained *Cleopatra* apples for chemical analysis from Mr Lang's orchard at Harcourt, where they had received only natural rainfall, for comparison with others which had been irrigated several times. At Campbell's Creek there was an interesting experiment with pears in which grafts had been taken from trees affected by bitter pit and worked onto trees free from it and *vice versa*. McAlpine found that some of the grafts had produced fruit, and the results were being carefully noted. The pips obtained from the cross between *Cleopatra* and *Yates* had germinated and the young seedlings had been photographed. In order to test the effect on the fruit of checking [interrupting or slowing] transpiration, an apple tree was selected with a light crop of fruit and heavy foliage. A bearing branch was cut into the wood with a saw and another had the leaves removed. He inspected the *Cleopatra* trees at Deepdene where zinc bands were being used and pruning experiments carried out.

March 4th 1914 At Burnley Gardens, the numerous experiments had been inspected and the young trees on the various rootstocks were bearing fruit. The transpiration of growing apple had not hitherto been determined, and this was being done with *Yates* growing under usual practice orchard conditions. The *Williams' Bon Chretien* pear tree at Box Hill, which had borne good fruit for 3 years since being 'rung', died after producing an abundance of blossom and a good fruit set. At Deepdene near Melbourne, important pruning experiments had been conducted over three seasons, but owing to the land being set aside for building sites, the trees were to be uprooted. In Tasmania, the experiments with *Cleopatra* apple trees were inspected and arrangements made for recording the results of different methods of pruning on the development of bitter pit. At Rheola, 6 miles from Inglewood in Victoria, there was considerable reduction in bitter pit in apples on treated trees. An 8-year-old *Dunn's Favourite* tree had produced the heaviest known apple, weighing 34 oz [i.e. 2.125 lb=0.966 kg]. *Cleopatra* apples from experimental plots at Stanthorpe in Queensland and Yanco, in the Murrumbidgee Irrigation Area of NSW, were forwarded [to Melbourne] for the purpose of storing them under different conditions.

April 4th 1914 March was spent mostly recording and analysing the results of various experiments relating to bitter pit. McAlpine made his annual visit to Western Australia where a complete series of manurial experiments was under way. Since there was a good crop of fruit on the *Cleopatra* trees, interesting comparisons could be made between the 16 different plots. The Tasmanian experiments were also recorded, pruning experiments being conducted both in the Tamar River Valley, north of Launceston, and at New Norfolk, 12 miles north west of Hobart. At Mr Edgell's orchard, Bathurst, NSW, different methods of pruning were carried out, but the experiment was largely discounted as a result of the most violent gale that ever struck the district, knocking off about one half the crop. At Burnley Gardens and at Deepdene, 6 miles east of Melbourne CBD, the results of pruning experiments were also recorded. Experiments were in progress at Diamond Creek, 18 miles north east of the CBD, testing the effect of sap flow retardation on the development of bitter pit. This was achieved by breaking fruit-bearing branches early in the season, the effect of which was markedly beneficial. Comparisons were being made between the keeping qualities of fruit from irrigated and non-irrigated trees with regard to keeping quality and the development of bitter pit. One case each of *Cleopatra* apples from Yanco and Stanthorpe respectively had been placed in cold storage, while similar cases were kept in the fruit room under ordinary conditions.

May 8th 1914 During April, McAlpine was engaged mostly in collecting results from experiments at Burnley Gardens, Cowan's Orchard at Bacchus Marsh and Hatfield's orchard at Box Hill. At Burnley the apples were gathered and recorded from the 56 varieties on various rootstocks planted in the bird-proof enclosure. Results were also gathered from other experiments. The transpiration of developing *Yates* apples was determined once a week by weighing the calcium chlorite tubes to measure their weight gains through moisture absorption. At Cowan's orchard the results of irrigation experiments were determined, and at Hatfield's orchard the manurial tests for the third season showed the proportion of pit associated with the different fertilisers. At Toomuc Valley orchard near Pakenham, experiments comparing different systems of pruning *Champion* apple trees were yielding very instructive results. A number of trees in one row had had their limbs cut back for grafting, with the exception of two branches left of the original tree, which bore a heavy crop of 92 % pitted apples, while an adjoining tree of the same variety, with only the leaders topped, produced only an occasional pitted apple. A number of specimens were forwarded from various districts for determination. The story of the variation in the incidence of bitter pit on the two *Champion* trees was a device McAlpine used to give some interesting information in his monthly reports, based on casual observations, while avoiding the provision of detailed results of experiments.

June 3rd 1914 During May, McAlpine was analysing the results of the last season's experiments and preparing for the coming season. He visited Mead, about 20 miles from South Kerang in the Cohuna district where apple trees were being destroyed by salting of the soil and found no pit in *Cleopatra* apples, while in another variety, *Senator*, on unaffected land there was as much as 50 % pit. *Cleopatra* apples were x-rayed to determine the absence of bitter pit even in its incipient stages. The apples were then kept along with non-x-rayed apples to watch for developments and to see if x-rays favoured or hindered the development of the bitter pit after exposure. At the Interstate Fruit Show in Hobart, the largest and most representative collection of apples ever brought together in Australia was displayed, and advantage was taken of the opportunity to record the varieties which pitted. McAlpine obtained the largest apples on display, including a *Prince Alfred* weighing 1 lb 6 oz [625 g] and an *Annie Elizabeth* 1 lb 4.5 oz [582 g], which were kept to see if they developed pit. Cases of *Cleopatra* apples were forwarded from Yanco in NSW and from Stanthorpe in Queensland. One half was kept in cold storage for 2 months and the other half in an ordinary fruit room. It was noticeable that the fruit from the cold store came out as green as it was when put in, while the other had turned yellow and ripened. The transpiration experiment continued until the apples fell from the tree on May 30th.

July 4th 1914 During June, McAlpine was mostly planning and arranging experiments for the coming season. Manures had been applied to and ploughed into the plots at Box Hill and peas sown for green manuring. The zinc-banded trees at Deepdene were examined and the apples from the plots kept in cold storage. He visited Diamond Creek where several *Cleopatra* trees had had their branches broken so that they remained attached only by a portion of the broken wood. More flowers and more fruit were eventually produced on the broken branches than on the whole. The effect on bitter pit remained to be seen. He visited Ashburton 7 miles south east of the CBD where 8.5 % of *Josephine* pears were affected with bitter pit. A small box of 25 *Jonathan* apples sent from Somerville, 38 miles SE of the CBD on the Mornington Peninsula in March, and similar to those exported, were stored until examined on 16th June, when 64 % were pitted, and 36 % were clean. The *Annie Elizabeth* tree at Burnley Gardens with various branches 'ringed' was examined with respect to the callus. In no instance had the callus enclosed the wound and in every case it was most developed on the upper or outer margin. A young tree, a cross between *Cleopatra* and *Yates*, was planted out, as well as several *Cleopatra* budded on pear and varieties of apples on their own roots.

August 4th 1914 McAlpine visited Tasmania during July in connection with the pruning experiments at New Norfolk and on the Tamar, and also arranged for testing the effect of shelter on the development of bitter pit in apples. The experimental

Cleopatra trees at Deepdene were pruned and gypsum applied in order to break up the stiff clay. The laterals on 14 varieties of apple trees were examined and the fruit spurs and sub-laterals on each year's wood noted. Arrangements were made to have each one netted so that any windfalls would be accounted for. The pruning at Burnley Gardens was also completed and the 54 young trees in the screen attended to. A case of *Cleopatra* apples kept in cool store was examined and only two pitted apples, each with only one pit, were found. The pits may have been present when the apples were placed in store. The large 1 lb 3 oz *Annie Elizabeth* apple [582 g] collected in May had developed 30–40 pits, which were distinctly visible on the surface. When Mr Lang of Harcourt first picked the apple he especially noted that there was no pit evident. An investigation was attempted to determine whether such pitting was present in an incipient form when apples were still attached to the tree, or whether it developed after picking.

13.2 The Third Progress Report

The third report (McAlpine 1914) dealt with control of bitter pit in the growing fruit. One aspect he examined was whether the age of trees influenced the incidence of bitter pit. The Chairman of Commissioners during the Dominion Royal Commission into bitter pit, Sir Ryder Haggard[A1], had concluded that whereas in Britain the fertility of soil gave out before the apple trees did, in Australia the productive life of apple trees diminished before soil fertility was exhausted. McAlpine rebuffed this view based on his own experience, citing apple and pear trees in an orchard in the Bathurst district, which were 89 years old, and continuing to bear as well as they ever did. He also noted several orchards in Victoria where 50-year-old trees were still highly productive and provided photos of them.

In attempting to get a better understanding of the impact of water translocation and transpiration on bitter pit during the development of fruit, a somewhat elaborate experiment was established to measure the transpiration rates from fruit during the critical months of the Victorian growing season, November, December and January, while ensuring optimal water availability. While providing him with interesting data, it did not advance his understanding of the causal processes. He still considered the cause was related to excessive and irregular sap flow. Several procedures were described for reducing the flow of sap in whole trees or in one or more main limbs of a tree in order to examine the impact on fruit development and the incidence of bitter pit. He conducted experiments using X-rays to test the possibility of determining the onset of bitter pit in highly susceptible varieties earlier than by microscopic or normally visible means. The use of X-rays did not prove to be useful in this regard. He also reported experiments determining the importance of low temperatures in maintaining the health of apples in storage, finding that healthy apples stored at 31 °F to 32 °F [−1 °C to 0 °C] were in the same state of health when removed from store many weeks later, as when they went in.

Manurial trials established in earlier years were continued and new ones established. Several of the treatments included calcium-rich components such as bone dust or ground limestone, but results between treatments and comparisons between the results of the same treatments in different locations provided McAlpine with no indication that calcium was a key to the cause of bitter pit. In fact, he saw the chief benefits of using lime as hastening the breakdown of green manure crops and in raising soil pH. Manurial treatments that gave a large lift in production in the first year mostly fell back in years two and three. More commonly than not, however, the heavier crops were less affected by bitter pit, with apple sizes more even and not large. One interesting observation related to potential benefits of rotation. However, recognising the impracticability of short-term rotations in orchard practice, McAlpine proposed that experiments examining potential benefits of rotating the manurial treatments used in orchard practice could be worthwhile. In this third report he referred to some trials running at the Yanco Irrigation Research establishment in the Murrumbidgee Irrigation Area. He also provided maps of the main apple growing areas of each state other than Western Australia.

References

Crathers W (1913) Letter written to W. Laidlaw [18-10-13], from the Secretary of the Advisory committee regarding the printing of the second progress report and foreshadowing an Advisory Committee meeting for mid to late November 1913

McAlpine D (1913) Bitter pit investigations. Second Progress Report – the cause of bitter pit: its contributing factors, together with an investigation of susceptibility and immunity in apple varieties. Government Printer, Melbourne, 224 pp, bibliography 96 pp, and 61 plates & maps

McAlpine D (1914) Bitter pit investigations. Third Progress Report – the control of bitter pit in the growing fruit. Government Printer, Melbourne, 176 pp and 77 figures

Chapter 14
The Fourth Year of the Investigation 1914–1915

Political unrest in Europe had been increasing since the beginning of the twentieth century, with the first decade witnessing increasing turbulence, distraction and uncertainty. With the outbreak of World War 1 on 28th July 1914, the focus of key people in government and the public service was not necessarily on McAlpine's investigation and added to his workload. The outbreak of war coincided with the start of the BAAS conference. McAlpine anticipated an influx of friends and overseas colleagues from July to October to attend the British Association Congress during August and September at Melbourne University (Fig. 14.1). This coincided with the busy period of setting up field trials and experiments for 1914/1915 without the assistance of numbers of key field workers who had help establish and maintain experiments in each State. The following 3–4 years were a period of increasing difficulty and stress.

14.1 Royal Commission

By 1914 the Victorian Government had recognised the need for an overall review of the horticultural industry because of the extent and pace of its expansion. Thus, on 10th February 1915 a Royal Commission on Fruit, Vegetables and Jam was established, which was submitted to Government on 14th October 1915 (Anon. 1915). The commissioners were required to report on and make recommendations regarding the production, packing, marketing, preservation, distribution, cold and other storage and wholesale and retail prices of fruit and vegetables. They were especially charged with investigating all financial aspects of businesses and trading agreements, understandings sanctioned, methods adopted, and practices followed by growers, traders, manufacturers and others, regarding Victoria's pecuniary interests in all aspects of the industry defined above. The commissioners interviewed McAlpine and five other witnesses regarding bitter pit, and among their various

D.G. Parbery, *Daniel McAlpine and The Bitter Pit*,
DOI 10.1007/978-3-319-09552-3_14

Fig. 14.1 Group at the British Association for the Advancement of Science Congress, Melbourne 1914. Those in the photograph referred to in various chapters are: Back Row L to R 2nd Isabel Cookson, 3rd Ethel McLennon, 12th E.E. Pescott, 13th Daniel McAlpine, 14th C.C. Brittlebank. Centre Row 4th A.J. Ewart. Front Row 2nd C.A. Topp (Permission, Special Collections, University of Melbourne Archives, Baillieu Library)

recommendations were some that either arose from McAlpine's findings or related to them.

The first witness was Professor Ewart who maintained that bitter pit was caused by minute traces of poisons, not necessarily sprays, but present in many soils and taken up through the roots then translocated to various points in the branches where they accumulated. He had no suggestions as to how to avoid the problem.

McAlpine rebutted claims that bitter bit was caused by microorganisms, lead arsenate or other toxins. He drew attention to the collapse of cells around vein endings, the fact that such tissue contained less water than cells in healthy tissue and the fact that bitter pit was always worst in seasons when dry periods ended in a shower of rain. He believed that under these circumstances the fruit expanded quickly but that the extension of the vascular networks failed to keep up with this expansion, so that some cells were then cut off from essential resources, died and shrank, causing pits. He explained the need to avoid applying nitrogen-rich manure such as horse manure, his preference being green manure crops, and he noted the benefit of cultivating orchards after rain, which helped to conserve soil moisture and reduced pit. Good drainage of orchards also tended to reduce pit. He emphasised the need for light pruning and the need to select apple varieties for different localities that were not prone to bitter pit. He spoke of the need to pick only mature fruit and, immediately after harvest until marketed, to hold it in cool storage at 32 °F. He emphasised the importance of selecting the best rootstocks for the various apple varieties grafted on to them.

Of the four following witnesses, the evidence of G.P. Darnell-Smith, Biologist, and W.W. Froggatt, Entomologist, each from the NSW Department of Agriculture, strongly supported McAlpine. George Quinn from South Australia said he doubted there was any evidence to support Professor Ewart's poison theory, but at the same time he was not convinced that McAlpine had given the right cause. The final witness, Henry M. Nicholls, the Tasmanian Government's microbiologist, proposed a novel explanation of the cause, which he attributed to the yeast phase of *Coniothecium chomatosporium*, which he claimed was almost universally present in orchards. He said the yeast developed in the vascular system and was transported to vein endings where it produced minute foci of fermentation thereby killing adjacent cells, which formed the pits.

Of the 31 recommendations submitted to Government, four were either specific to bitter pit or relevant to it (Anon. 1915). In their 23rd recommendation they stated:

> We were impressed with the earnestness and energy which these gentlemen have brought to their investigations, and consider that a stage has now been reached when they might all be brought together by their respective Governments to confer upon their work, results and opinions. Your Commissioners, therefore, strongly recommend that this course be adopted as soon as practicable, and before Mr McAlpine's final report is published.
>
> In the meantime we think that the fruit growers should be made fully acquainted with definitely decided results on the subject of this most baneful disease – of Bitter Pit, as they are obtained. For instance, no time should be lost in advising them in respect of the varieties of apples which are immune from the disease, proper pruning, manuring, drainage, cultivation and stocks, so that the ravages of the disease may be mitigated, if not entirely prevented.

The commissioners visited many apple-growing areas in Victoria, as well as Sydney and Hawkesbury in NSW, Launceston and Hobart in Tasmania, and Adelaide. In addition they visited storage facilities at several points including cool stores at ports in Melbourne and Sydney. They held 52 meetings, interviewed 135 witnesses and received numerous submissions, including one from the Agent-General in London and another from the Canadian Government. They submitted 31 recommendations to the government, which resulted in many changes, some of which were based either on McAlpine's evidence or his first three reports. In addition to Recommendation 23, four others are noteworthy and several others are of interest in terms of economic history.

In Recommendation 11, the Commissioners called for better arrangements at Covent Garden, London, for the handling of Australian produce. Not only storage facilities needed improvement but also payments needed to be made and recorded in a structured manner and properly accounted for.

Two recommendations dealt with railway services. Number 8 recommended that the railways adopted the Post Master General's procedures of delivering goods to the door so that fruit and vegetables sent by rail did not languish in goods yards or station buildings. Recommendation 22 dealt more specifically with the way in which fruit and vegetables were handled so as to avoid bruising in transit. They advocated installing railway sidings at all fruit growing centres; the availability of much greater numbers of iced louvered vans; and the introduction of a form of 'container' van body capable of being lifted from under-frames of one gauge to those of another at break-of-gauge points so as to minimise handling. They went as far as recommending the abolition of loose-chain coupling of railway vehicles, and the avoidance of throwing and dropping boxes of fruit, which resulted in severe bruising. McAlpine had earlier stressed the importance of handling apples and pears as gently as possible to avoid damage. Recommendation 30 stressed that diseased fruit should not be allowed in or out of the country.

Many of these recommendations were put in place. Blackburn, about 12 miles [20 Km] east of Melbourne was a typical fruit growing district. Now an outer suburb, but then a village, it was surrounded by numerous large apple orchards. Settled in the 1840s, by the 1880s the local fruit growers' cooperative built a large fruit store with access to a railway siding. In 1918 the cooperative installed a cool store and began exporting large volumes of apples to Britain. Interesting too, the Victorian Railways began investigating the possible of introducting buckeye couplings to replaced loose chain couplings between goods wagons at about the same time, which they did in 1923 (Gregory 2014).

14.2 The Fourth Progress Report

In the fourth, and what he believed would be his final report, McAlpine (1915a) provided additional data from manurial, pruning and other trials, and observed that 4 years was not long enough to allow many of the experiments to yield useful results. Such experiments included those established to determine the value of using different root-stock/scion combinations, selection and breeding for resistance to bitter pit and the continuation of manurial trials, which he believed needed to run for 10 years or more. He summarised the outcome of his 4-year investigation in the following 13 points, which, if followed, would not avoid bitter pit totally but would reduce its incidence to manageable proportions:

1. Ensure that apples are picked and packed as carefully as eggs to avoid bruising which initiates deterioration;
2. Do not allow trees to bear too early, as they frequently prove very susceptible to bitter pit;

3. Ensure that once picked fruit goes as soon as possible in to cool store. Surprisingly, McAlpine did not add that only fully mature fruit should be picked and stored at 31–32 °F, which he had emphasised in other places;
4. He noted that the freezing point of apple juice was 28 °F to 29 °F [−2.2 °C to −1.7 °C];
5. Because of varietal differences highly susceptible varieties such a *Cleopatra*, grown in localities conducive to bitter pit, should be replaced with more resistant varieties;
6. Nitrogen-rich manures should be avoided;
7. Apply lime to hasten the breakdown of green manures and the liberation of nitrogen and small amounts of potash and phosphoric acid, and should be applied two-yearly;
8. Trees in well structured, fertile soils should not be manured [fertilised] until they begin to bear;
9. Poorly drained soils favour the development of bitter pit;
10. Avoid severe pruning which favours oversized fruit, which is more prone to bitter pit than average sized fruit;
11. Use irrigation or other means of artificial watering only to even out the available water supply through the growing season, and avoid over watering;
12. Maintain well-shaped trees to preserve uniformity of bearing and fruit size;
13. Pay attention to the best rootstock/scion combinations for the locality of the orchard.

Finally, McAlpine acknowledged the strong support he had received from Dr S.S. Cameron throughout the investigations, the assistance he had been given by Mr C.C. Brittlebank and Mr C. Wedge and the many orchardists and agricultural officers in the various states who had assisted with the trials and experiments. Among the latter he paid special tribute to Mr George Quinn, Horticultural Instructor in charge of the South Australian Experimental Orchard.

14.3 The Advisory Committee's Penultimate Report

During early September 1915 the Advisory Committee considered McAlpine's Fourth Progress Report. Committee Members clearly expressed disappointment over the lack of advance in finding the cause or causes of bitter pit (Cumpston 1915). The main reason for this was that they saw the primary objective of the investigation as ascertaining the cause of bitter pit. McAlpine had expressed the same sentiment with regard to diagnosing diseases caused by biotic agents (1910), but in the case of bitter pit he had given priority to seeking changes to the growing, storage and transport of fruit that would reduce the problem. The Advisory Committee acknowledged the vast amount of work McAlpine had undertaken, but made no comment about a reduction in the incidence of bitter pit as a result of it. They summarised 11 aspects of the factors McAlpine had identified as contributing to the development of bitter pit which involved sap flow, death of cell clusters near vein endings, rates of transpiration and seasonal variations, but said they did not agree with some of the conclusion drawn as they were based on inadequate data. They implied that there needed to be more data collected to give greater reliance to his recommendations regarding pruning, manuring and other orchard practices to reinforce the first of McAlpine's summarising statements that; 'it is by the control which the intelligent orchardist can exercise over the growth of the trees and the formation of the fruit that the disease can be kept within manageable limits'. His second statement was to the effect that the best long-term solution to bitter pit was the breeding of resistant varieties.

Later in the report, however, they did acknowledge benefits from the investigation. They congratulated the Commonwealth and State Governments for setting up the investigation. They noted that there were several phases of the investigation that had cleared away certain disabilities and noted the comprehensive review of previous literature and the full delineation of the vascular system in the First Progress Report. However, they were not convinced by the explanation given for the connection between the deaths of the cell clusters beneath the pits around vein endings. They referred to the potential value of Drs F. Lemon's and H. Lawrence's report that incipient pitting may be detected in apparently healthy apples, but cited no reference. At the time of writing his third report, McAlpine (1914) was unable to report a positive result from the use of x-ray to identify incipient pitting, as noted earlier in Chap. 13. By the time of the Advisory Committee's fourth meeting, however Lemon and Lawrence had picked fruit from a tree with a history of serious pitting. They x-rayed eight superficially sound apples and noted one that exhibited unusual clusters of cells beneath the skin. After a period of storage under ambient conditions it was the only one of the eight apples that developed bitter pit. The Advisory Committee acknowledged that Dr Rother's investigation needed to be taken further, and substantiated Wortmann's (1892) findings, and those of other early observers that there were critical periods during the development of fruit when it was most liable to develop bitter pit. The final point in their appraisal was to acknowledge the importance of McAlpine's discovery that storage at 30–32 °F arrested the development of incipient bitter pit, a finding of great value to commercial fruit growers and shippers alike. Their concluding recommendations on page 78 were:

1. That the importance of the subject justifies continuing the investigation.
2. That the investigation be limited to the cause and control of bitter pit.
3. That the Advisory Committee controls the work of the Investigator.
4. That the contributing Governments be recommended to invite McAlpine to continue the investigation.
5. That the work of the Investigator be confined as far as practicable to one State, and that Officers in other States continue to assist in carrying out experiments decided upon by the Advisory Committee.
6. That the Investigator should visit the contributing States in order to provide advice on work carried out by State Officers, at such times as may be determined by the Advisory Committee.

7. That in view of the fact that the scope of the work would be materially altered the maximum expenditure be fixed at £500 for the salary of the Investigator and £500 for expenses; but that the Investigator be permitted to accept other work at the discretion of the Committee.
8. That the full Committee be authorized to meet once every 6 months, and the members resident in Melbourne be authorised to meet quarterly.
9. That any agreement entered into between the controlling Ministers and the Investigator be subject to review at the end of every 12 months.

Concluding their report, the Advisory Committee added another recommendation, that the attention of the Commonwealth Government, as the body controlling the export trade, be advised as to the desirability of taking action necessary to ensure that fruit was cooled before being placed on board any vessel.

It seems evident that although the Advisory Committee was disappointed in aspects of McAlpine's investigation, they identified several positive outcomes and were prepared to recommend his employment for a further 12 months in the first instance. However, the committee wished to limit the extent of McAlpine's autonomy and to be authorised to direct the investigation.

It is notable that at some stage during 1915, possibly after Dr Rothera's untimely death[A1], his finding that there were no chemical differences between pitted and non-pitted apples, and no evidence of lead or arsenic present in the latter (Beidahl and Rothera 1915), was presented to the Royal Society in refutation of Professor Ewart's (1915) assertions. Had Dr. Rothera been aware of the importance of calcium as a plant nutrient when he made chemical comparisons of pitted with non-pitted apples (Chap. 12.1), it may have led to a much more satisfying conclusion to the investigation for McAlpine than was his fate!

14.4 McAlpine's Letter to the Premier

By the end of 1915, McAlpine had stated in the conclusions to his Fourth Progress Report that he had taken the investigation as far is he could and was not anticipating an extension of it, but rather he was looking forward to resuming his old post as Vegetable Pathologist, in which role he would have the oversight of the breeding experiments set up during the investigation. Thus on 12th November 1915 McAlpine (1915b) wrote to the Premier as follows.

> 'The Hon. Sir Alexander J. Peacock, Premier, Victoria
>
> I have much pleasure in forwarding herewith my latest Report on Bitter Pit in Apples, a disease which is prevalent in Europe, and America, Asia and Africa, as well as Australia. It is a disease that has puzzled scientist all over the world and it has been known as the 'Mystery' of Bitter Pit.
>
> You will be pleased to learn that the problem has now been solved, as far as our present knowledge goes, in the Commonwealth of Australia. You will observe that the Cause of Bitter Pit has been determined, as well as its Control, both while the fruit is in store and while growing on the tree.

The over-seas export of Apples is now established on a satisfactory footing, since I have shown that if the fruit is carried at a uniform temperature of 30–32 degrees Fahrenheit the development of Bitter Pit is prevented and the over-ripening of fruit is arrested.

The Practical Applications on p. 78 give a summary of the essential points to be attended to in connection with successful commercial apple growing.

I remain,

Yours faithfully,
D. McAlpine.

References

Anon. (1915) Report from the Royal Commission on fruit, vegetables and jam. Government Printer, Melbourne, 54 pp

Breidahl HG, Rothera ACH (1915) Bitter pit and sensitivity of apples to poison. An answer to Prof. A.J. Ewart. Proc R Soc Vic 27(2):191–197

Cumpston JHL (1915) Report [10-9-15] submitted by the Advisory Committee on the bitter pit investigation during 1914–1915

Ewart AJ (1915) On bitter pit and the sensitivity of apples to poisons. Proc R Soc Vic 26:228–242

Gregory DM (2014) Email [25-6-'14] sent to D.G.Parbery

McAlpine D (1910) Twenty years of plant pathology in Australia. Report of the Vegetable Pathologist for 1907–1910. Department of Agriculture, Victoria, pp 56–61

McAlpine D (1914) Bitter pit investigations. Third Progress Report – the control of bitter pit in the growing fruit. Government Printer, Melbourne, 176 pp and 77 figures

McAlpine D (1915a) Bitter pit investigations. Fourth Progress Report – the experimental results in their relation to bitter pit and a general summary of the investigation. Government Printer, Melbourne, 178 pp

McAlpine D (1915b) Letter [12-11-'15] to Sir Alexander Peacock, telling him the investigation has gone as far as possible

Wortmann J (1892) Uber die sogenannten "Stippen" der Äpfel. Landwirtschaftliches Jahrbuche 21:663

Chapter 15
Fifth Year of the Investigation 1915–1916

15.1 The Uncertainties of 1915–1916

Three days after McAlpine (1915d) wrote to Premier Peacock stating he had taken the investigation as far as possible, Prime Minister Billy Hughes (1915) wrote to the Premier telling him the bitter pit investigation would be extended for a further 12 months. Six days later, *The Age* reported (Anon. 1915) that McAlpine hoped the investigation would be extended so as to enable long-term breeding experiments to continue. The article also referred to the publication of his Fourth Progress Report (McAlpine 1915c), which it said, like all others, was a substantial volume, copiously illustrated and expensively produced. Shortly after this, Dr Cumpston (1915b) wrote to the Premier telling him the Government printer had been instructed to forward the usual number of copies of the report to the Agricultural Departments of each State. Thus the Prime Minister and Premiers should have been aware of the advice to extend it but under different conditions.

Newspapers were aware of the public's interested in the investigation and published excerpts of McAlpine's reports about the effects of orchard management on the incidence of bitter pit. Two of the published tables, demonstrated this and apart from their general interest to growers should have been useful pointers as research directions for those interested in plant physiology. While Ewart fixed his mind on asserting his poison theory, McAlpine fixed his primarily on seeking the best cultural practices possible to help the orchardists, and this was where he directed his energies. While his approach did not satisfy the Advisory Committee, it greatly satisfied the struggling orchardists and others in the fruit trades worldwide. Industry mounted pressure for the investigation to be extended. The first trial of public interest related to nutrient levels, tree vigour and the incidence of bitter pit (Table 15.1). To one plot of trees [Treatment 1]

D.G. Parbery, *Daniel McAlpine and The Bitter Pit*,
DOI 10.1007/978-3-319-09552-3_15

Table 15.1 Results from McAlpine's manorial trial (*The Argus* 11-3-1915)

Treatment	% Pitted fruit	% Windfalls	% Sound
1. Nutrient rich	33	51	16
2. Nutrient poor	18	26	56

Table 15.2 Results from McAlpine's pruning trials (*The Argus* 20-4-1916)

Pruning regime	Clean fruit	Pitted fruit	Total yield	% Pitting
None	511.5	31.0	542.5	5.17
Leaders removed	788.5	34.5	822.5	4.16
Light	583.0	39.0	622.0	6.27
Severe	480.5	135.5	616.0	22.36

McAlpine had added the usual application of farmyard manure plus 105 lb per tree of superphosphate, while to the second [Treatment 2] he added no farmyard manure but 50 lb per tree of superphosphate.

There is little doubt that McAlpine would have used this information in meetings with orchardists and their advisors to confirm the message he had been putting out for some time, especially with regard to high levels of nitrogenous fertilisers increasing the incidence of bitter pit. He had already demonstrated that by applying high levels of rich farmyard manure to orchards of *Cleopatra* apples, one of the most susceptible varieties, the incidence of bitter pit was often as high as 70 % and occasionally 95 %. While avoiding the overuse of nitrogen-rich fertiliser more than trebled the proportion of sound fruit. There was still clearly much room for improvement.

The second trial demonstrated the impact of pruning levels on bitter pit (Table 15.2). He had noted that just as the application of high levels of nitrogen favoured high levels of leaf growth so did certain pruning regimes. He had noted that where there was abundant new leaf production in competition with the developing fruit, the incidence of pit increased.

Other results that caught the attention of the newspapers were those from a pruning trial. In early 1916 *The Argus* published the following information: Pruning trials began in 1911; *Cleopatra* scions grafted on to *Northern Spy* rootstock were begun in 1911; the 1915/16 season pruning was done on 2nd August 1915; the fruit was picked on 3rd and 4th April 1916, and the yield of fruit and the levels of bitter pit were recorded (Table 15.2).

Later that year, in its 25th October issue, *The Argus* added that late pruning, just as sap flow began, gave even better results in reducing bitter pit, so that not only the level of pruning but also the timing was important.

McAlpine's 4-year contract with the Advisory Committee was due to expire at the end of 1915, and during the year he was looking to the future, expecting to return to his old post as Vegetable Pathologist. An extension of this expectation was that as such he would be able to continue his pruning, grafting and breeding experiments without interruption. On 28th July he wrote to The Hon. William Hutchinson (McAlpine 1915a), then Victorian Minister for Water Supply and Agriculture, explaining that his contract with the Commonwealth and State Governments would soon expire and asking what his position was in regard to the State service, pointing out that it was 25 years since his appointment in May 1890. He received no response. According to family sources (Wedge 1984) he believed his gentleman's agreement with the earlier Minister George Graham would be honoured at the end of the bitter pit contract and he would return to his old position. Sadly that was not the case but it appears that he was not made aware of it until early in 1917.

In the lead-up to the bitter pit investigation it was evident that McAlpine expected to continue his study of Australian fungi and plant diseases after the bitter pit investigation finished. A large part of the reason for accepting to lead the investigation was predicated on the basis that he would return to the Victorian Department of Agriculture at the end of it. During the first 2 years of the investigation, growing misgivings over his ability emerged within the Advisory Committee, mostly through the malign influenced of Professor Ewart. By the end of the third year of the investigation, McAlpine's relationship with the Advisory Committee had frayed. For example, in April 1915, McAlpine (1915a) wrote to The Hon. W. Hutchison complaining that he had been misquoted in refusing to provide the chairman of the Advisory Committee with a preview of the conclusions to be drawn from his investigation. He explained that this was not possible until the results of experiments, still to be gathered, were available. He may also have been concerned that they would be leaked to Professor Ewart, giving him further opportunity to distort the judgment of members of the Advisory Committee. On 2nd December 1915 his fourth report (McAlpine 1915c) was circulated to the Prime Minister, each State Premier and all Ministers of Agriculture.

15.2 Extension into a Fifth Year

In February 1916, the Prime Minister wrote to the Premier of Victoria, Sir Alexander Peacock (Hughes 1916a) inquiring whether the Commonwealth Advisory Committee on Bitter Pit would advise him on the need to extend the bitter pit investigation. The advice had been handed down in the Report of the Advisory Committee dated 10th September 1915 (Cumpston 1915a) but was somewhat ambivalent. The report concluded that while those conversant with other studies on bitter pit could not fail to be impressed with the vast amount of work done by Mr. McAlpine, the Committee was disappointed in being obliged to state that little substantial advance had been made in its scientific knowledge of the cause of the disease. The committee then distanced itself from any implied failure by placing on record their inability to direct it. They pointed out that they had no authority to control the course of the investigation or offer criticism, which in fact they had done in more than one report. They said, however, that they disagreed with some of the conclusions reached in the reports because they were framed on data the Committee regarded as inadequate. They agreed, however, that there would be merit in extending the pot trials and breeding program that had been set up at Burnley (Cumpston

1915a). This of course could have been done without there being a nationwide effort and it is what McAlpine was hoping for.

Prime Minister Hughes[A1] was heavily involved in the war effort as a member of the Imperial War Cabinet which met in London. It coincided with a serious escalation of World War 1 and horrendous slaughter of Australian troops in the battle for Fromelles and other theatres on the western front (McMullen 2002). Hughes was working hard to raise support for them. Similarly increasing workloads and concerns affected other politicians and bureaucrats and possibly distracted attention from communications over bitter pit between the Commonwealth and States. Charles Brittlebank's only son and a nephew were among the many killed in France. Communication between the Advisory body and McAlpine, was also breaking down. In consequence the period between the end of the 1914/'15 season and start of the next was one of confusion, distraction and disarray. The Hon. F.W. Hagelthorn (1915), newly appointed Minister for Agriculture, advised Premier Watt to recommend that the Prime Minister extend the investigation for a further 2 years. Prime Minister Hughes extended it for 12 months, but having given his approval for this in mid November (Hughes 1915), he enquired again the following February (Hughes 1916a) if the Advisory Committee supported the need to extend it. Evidently he was told it did, because on 18th March he wrote (Hughes 1916b) that the States had agreed to continue but with McAlpine's salary reduced to £850 with expenses fixed at £500 per annum. On the same day the Hon. G.F. Pearce (1916a), who was Acting Prime Minister while Hughes was in London on Imperial War Cabinet business, wrote a similar letter advising the make up of McAlpine's salary for 1916 as follows:

Commonwealth	£500
Vic., Qld & WA each £83/6/8	£250
NSW, SA & Tas each £41/13/4	£125

The outcome of negotiations noted by Pearse (1916a) was that Victoria, Queensland and Western Australian agreed to the extension, at a salary of £1,000 per annum plus £500 expenses; New South Wales and South Australia supported the proposal but at a reduced salary of £500 p.a. plus £500 for expenses, while Tasmania expressed its wish to withdraw. On 7th June, the Commonwealth wrote to New South Wales, South Australia and Tasmania asking them to reconsider. This resulted in Tasmania coming in on the same basis as New South Wales and South Australia. As a consequence for the fifth year, McAlpine's salary was set at £875. However, McAlpine was not actually paid that amount, because in a letter Acting Prime Minister Pearce (1916b) advised that because four States had reconsidered their levels of support McAlpine's salary for 1916 would be £750; the Commonwealth contributing £500 and the States £41/13/4 each, with contingency expenditure limited to £500. The arrangement took effect from 1st January 1916, and the position was to be reviewed at the end of 1916. This was a retrospective decision made 10 months after the fifth year should have begun in August 1915. McAlpine's reaction to the reduced salary is unknown and neither is the date from which the fifth year was authorised to begin. Possibly he kept working in the expectation that he would be granted a fifth year on the basis that the Advisory Committee had recommended it and Hagelthorn supported extending it to the end of 1917.

In view of the fact that the Advisory Committee met to review his fifth report on 27th October 1916, it is not surprising that achievements during the final year were limited. The extension was not approved until well past the start of the 1915/‘16 season and there may not have been much enthusiasm for it in some quarters. The Advisory Committee was not strongly supportive, and McAlpine (1915d) had written to Sir Alexander Peacock in November 1915 saying he was finished. The content of the Fifth Progress Report essentially confirms that little was achieved in the troubled time available. It contains updates of data from various trials, and some miscellaneous observations of interest to the horticultural industries and especially to those responsible for continuing aspects of the investigation on State-sponsored bases. McAlpine provided photos of a large mature apple tree of the *Rymer* variety, which had produce 56.5 cases of apples in the 1915/1916 season. As a case of apples weighed 42 lb, this amounted to 2,543 lb [1,156 kg] of fruit. Of these 40.5 lb were picked and marketed, the rest were windfalls. He reported that because of its proneness to bitter pit, many orchardists were replacing the variety *Cleopatra* with less bitter pit-prone varieties and that one orchardist in Deepdene, a Melbourne suburb today, had cutback his *Cleopatra* trees and grafted them with improved varieties. He noted that in the USA, orchardists were about to embark on a similar operation, replacing the susceptible variety *Baldwin*, which had been as popular in North America as *Cleopatra* was in Australia. He also reported that the Director of the Royal Gardens, Kew, had sent him an assignment of crab apple stock from which it is evident he intended to develop a breeding program aimed at producing bitter pit resistant varieties following his expected return to the Victorian Department of Agriculture.

As far as the conclusion about how best to manage orchards to minimise the incidence of Bitter Pit was concerned, there was no advance beyond the conclusions in his Fourth Progress Report, or what was set down in the report of the 1915 Royal Commission into the horticultural industries which are summed up in his final publication in *Phytopathology* (McAlpine 1921). After the publication of his final report, he made a brief statement on his conclusion about the cause of bitter pit. He said that although prior to the fifth report he had considered that it was the concentration of cell sap in apple tissue that caused the death of cells beneath the pits, he had changed his view and at the end of 1916 believed it was excessive water pressure in apple tissue that lead to the rupture and consequent death of cells which was responsible, as proposed by Evans (1909).

15.3 The Final Advisory Committee Report

On 27th October 1916, the same day as the Advisory Committee met to formulate recommendations to the Commonwealth and State Governments, McAlpine advised the Committee that he had taken the investigation as far as possible, that he could see no further benefit in continuing on the same basis, and that this would be his final report as there was no longer any need for him to be involved. He did, however, see value in further investigation of the influence of different rootstocks on the incidence of bitter pit in grafted trees. He suggested that more or less immediate benefits would most likely flow from a broad study of graft combinations, and that he saw potential benefits from what he called an investigation into artificial evolution. His basic idea was that the crab apple, said to be the progenitor of domesticated apple varieties, never exhibited bitter pit. He expanded a little on this hypothesis in the final paragraph of his summary paper on bitter pit, possibly invited by the editor of *Phytopathology* (McAlpine 1921). In 1920, however, he submitted an article entitled *Evolution and its Practical Application* to a publisher in New York, but it was never published and was returned to the family after his death.

The conclusions of the Advisory Committee drawn from the October 1916 meeting were that:

(i) The investigation should not continue on the same basis;
(ii) The States should organise their own investigations, esp. Qld, NSW & Victoria;
(iii) The States should report annually to the Chairman of The Advisory Committee;
(iv) The Advisory Committee should report to the Commonwealth and State Governments each 12 months; and
(v) The Commonwealth should allocate £500 per annum to fund the ongoing work of the Committee.

On 13th December 1916, Prime Minister Hughes (1916c) wrote advising that it was proposed to discontinue the investigation as from 31st December 1916. The Victorian Minister for Agriculture, however, recommended that the Victorian Premier should advise the Prime Minister that while it agreed with most of the recommendation in the final Advisory Committee report, it would not adopt recommendations (iii) and (iv) without knowing their financial implications.

References

Anon. (1915) Article [18-11-'15] in *The Age Newspaper* reporting that McAlpine's Fourth Progress Report had been published and like all others it was a substantial volume, copiously illustrated and expensively produced

Cumpston JHL (1915a) Report [10-9-15] submitted by the Advisory Committee on the bitter pit investigation during 1914–1915

Cumpston JHL (1915b) Letter to Dr. Cameron [1-12-'15] reporting his instruction to the Government Printer to distribute copies of the Fourth Progress Report to each State

Evans IBP (1909) Bitter pit of the apple. Transvaal Department of Agriculture, Technical Bulletin No. 1, Pretoria

Hagelthorn FW (1915) Hagelthorn wrote [13-11-'15] advising Premier Watt to recommend that the Prime Minister extend the investigation for two years

Hughes, The Hon. WM (1915) Letter [15-11-'15] to the Premier of Victoria telling him the bitter pit investigation was extended for twelve months

Hughes, Hon. WM (1916a) Wrote (Feb. 1916) to the Premier of Victoria, inquiring whether Advisory Committee supported the need to extend the bitter pit investigation

Hughes, The Hon. WM (1916b) Letter [18-3-'16] to Victorian Premier stating all States had agreed to extend the investigation at the salary of £1,000 p.a. but not all States agreed to this and after negotiations the offer was made and accepted by McAlpine at £875 per annum plus £500 expenses

Hughes, The Hon. WM (1916c) Circular [13-12-'16] sent to Premier of Victoria regarding the discontinuation of the Bitter Pit Investigation

McAlpine D (1915a) Letter written to Minister Hutchison [21-4-'15] explaining that the Chairman of the Advisory Committee had misquoted him

McAlpine D (1915b) wrote [28-7-'15] to Minister Hutchinson advising that his contract with the Commonwealth was about to expire and asking what his position was in the State service

McAlpine D (1915c) Bitter pit investigations. Fourth Progress Report – the experimental results in their relation to bitter pit and a general summary of the investigation. Government Printer, Melbourne, 178 pp

McAlpine D (1915d) Letter [12-11-'15] to Sir Alexander Peacock, telling him the investigation has gone as far as possible

McAlpine D (1921) Bitter pit on apples and pears. The latest results in preservation measures. Phytopathology 11:366–270

McMullen R (2002) Pompey Elliott. Scribe Publications Pty Ltd, Melbourne, 718 pp. ISBN 0908011 70 9

Pearce, Hon. GF (1916a) Letter sent to Premier of Victoria [18-3-'16] advising the make up of McAlpine's salary for 1916

Pearce, Hon. GF (1916b) Letter sent to Premier of Victoria [14-6-'16] revising the make up of McAlpine's salary for 1916

Wedge E (1984) Letters [4th to 7th May-1984] to D.G.Parbery, La Trobe Collection

Chapter 16
Appraisal of McAlpine's Success

16.1 Physiological Disorders

McAlpine (1910) had recognised that bitter pit was a physiological disorder prior to 1910, such problems not having been recognised previously in Australia. According to Ainsworth (1981), little was known about non-parasitic diseases prior to the twentieth century. He pointed out that Paul Soraurer[A1] was the first to draw particular attention to this group of what McAlpine referred to as 'Constitutional Disorders', and that by the time the third edition of Soraurer's *Handbuch der Pflanzenkrankheiten* (1909) was republished from 1905 to 1913 one volume was devoted solely to physiological disorders. Ainsworth drew attention to the generally limited space devoted to these diseases in texts on plant pathology, and made the point that plant pathologists generally disregard them, preferring to hand them over to plant physiologists or others. Ainsworth also drew attention to the lack of knowledge about the importance of minor and trace elements in plant nutrition before the 1920s so there was no base from which to study physiological disorders involving nutritional imbalances. Lack of boron in highly leached sandy soils was reported and confirmed as important for various crops between 1910 and 1923, and magnesium deficiency was identified as a cause of chlorosis in crops, including apples, growing in leached sandy soils. Over the next two decades, various experiments with boron and magnesium singly and in various combinations with major and minor elements were reported but usually with inconclusive of conflicting results (Bünemann 1972). McAlpine included both boron and magnesium in his manurial experiments during the bitter pit investigation, but without positive outcomes.

The benefit of lime in checking the incidence of bitter pit was noted at least as early as 1931 by Myer (Bünemann 1972), and 3 years later the influence of the presence of boron on calcium uptake was noted (Warrington 1934). In looking for chemical differences between healthy and pitted fruit in 1936, De Long (1936) made the seminal discovery that tissue from healthy fruit contained more calcium than that from pitted fruit. He also made the significant assumption that there was

D.G. Parbery, *Daniel McAlpine and The Bitter Pit*,
DOI 10.1007/978-3-319-09552-3_16

competition for calcium between actively growing shoots and young fruit. Unfortunately, the significance of these discoveries was not recognised at the time, so not taken further. Also in that year, Thiem (1936) reported that stippen was worst in orchards where there was a high water table and liquid manure and nitrogen were used. He recommended the use of slow release fertilisers and liming. McAlpine (1915) had earlier recommended the use of green manure, the equivalent to day of slow release fertilisers, especially in relation to nitrogen. Unfortunately neither De Long (1936) nor Thiem (1936) followed up on their discoveries, which were either ignored or overlooked by others. During the next 20 years, there was more interest in boron and magnesium uptake than in calcium, until Garman and Mathis (1956) published their benchmark paper. This opened the way to understanding the cause of bitter pit and provided the only substantial advance on McAlpine's recommendations for control since 1916. They found that the ratio of the balance between leaf and fruit calcium content was important, and that the availability of magnesium and potassium influenced it. They demonstrated that injection of calcium into orchard soil or the spraying of trees with solutions of calcium compounds prevented bitter pit. They also noted that high levels of soil nitrogen increased the uptake of potassium and magnesium more than it did calcium and was therefore detrimental. While McAlpine did not find evidence for the significance of calcium, he knew high soil nitrogen levels favoured shoot growth and increased pitting, and recommended avoiding it. Had he been aware of the nutritional importance of calcium he may have taken more notice of data from one of his small trials. For example, prior to the setting up of the national investigation, McAlpine (1910) ran a small trial with only two trees of a susceptible cultivar in each of two soil treatments and a control, which he called the 'check'. The results were shown in (Table 16.1).

These differences were not great and probably neither treatment was statistically different from the control. McAlpine may have considered it was the high level of phosphorus in dissolved bone that was important rather than the calcium. Had the lime made more of a difference it may have attracted his attention. Additional clues from the literature emerged too late for him.

While McAlpine was aware of the interest in the possible benefits of boron and magnesium, the most important findings were published well after his investigation finished and were too late to help him. He was working in a knowledge vacuum. This same knowledge vacuum compromised the Commonwealth Advisory Committee's ability to make objective assessments of McAlpine's progress. It is important to bear in mind that the knowledge was simply not available to help McAlpine advance beyond the endpoint he reached, or his judges to make a fair

Table 16.1 Results of a small trial run by McAlpine (1910)

Treatment	Bitter pit-free apples	Difference
Check	32 %	
Dissolved bone	45 %	13 %
Lime	36 %	3 %

assessment of it, had they been so inclined! In this context, his decision to concentrate on finding measures that would reduce or avoid the problem rather than focus on finding the cause was the best use of time.

16.2 Quantitative Evidence of Success

While there are plenty of data in Commonwealth and State annual reports on horticultural production it was often recorded in different ways in each State and at different times. Consequently, there are gaps where vital information is missing. For example, Commonwealth Reports included orchard acreages for each fruit type for every state except Victoria, which recorded only the area under gardens and orchards. Additionally, although the Victorian Minister for Agriculture and others referred at times to fluctuations in apple production from year to year, available information offered little help in identifying the cause. Thus unless quantitative assessments were made at the time it is difficult a century later to separate the impacts of drought, flood, late frosts, low bee numbers etc. on the incidence and severity of bitter pit. It is noted in Chap. 7 that 1904–1906 were bad seasons for bitter pit, but the big drops in apple production were seen in 1906 and 1908 (Table 16.2). The next dramatic drops were recorded in 1915, 1917 and 1919, when the first two bad harvests may have been largely drought related. Australia experienced a serious drought from 1911 through to 1916 such that in 1914 the Murray River ceased to flow for 6 months. In the bumper year of 1910–1911 the average yield per apple tree in Victoria was 48 lb weight of fruit, but in 1916–1917 it was 14 lb. Then again there was widespread drought during 1918–1920 coinciding with the pandemic of Spanish influenza in 1919. Added to this is the fact that yields recorded at harvest could not have taken account of the proportion of fruit, which developed bitter pit in storage or transit.

In the southern hemisphere, the growing season for pome fruits straddles the spring and early summer of one calendar year and summer into autumn of the following year as shown in Table 16.2, which records apple yields in Victoria before, during and after the period of McAlpine's investigation of bitter pit.

Table 16.2 Apple production in Victoria, 1904 to 1920 – before, during, and after the investigation

Before		During		After	
Year	Production	Year	Production	Year	Production
1904/1905	1,019,816				
1905/1906	578,700	1911/1912	1,330,961		
1906/1907	1,010,381	1912/1913	2,036,754	1917/1918	1,864,508
1907/1908	618,424	1913/1914	1,653,035	1918/1919	807,573
1908/1909	1,241,826	1914/1915	509,679	1919/1920	2,227,317
1909/1910	1,121,702	1915/1916	2,953,968		
1910/1911	1,667,272	1916/1917	617,929		
Average	**1,036,874**	**Average**	**1,517,054**	**Average**	**1,633,133**

1, Production is given in bushels. An Australian bushel of apples weighed 42 lb; 53.33. bushels = 1 ton [2,240 lb].

Prior to 1914, revenue was not included in all States' annual reports. In Victoria, however, they were. There, the annual average production of apples was recorded from 1904 and revenue from 1909. Over the seven seasons before the investigation began, average annual production was 1,036,874 bushels (Table 16.2). During the six seasons of the investigation, average annual production was 1,517,054 bushels, which was 46 % better than the 1904–1911 average, while the average for the following three seasons was 1,633,133 bushels, an increase of 57 % over the pre-investigation seasons but only a 7.7 % increase over the annual average of the investigation period. Australia wide, average yields rose from 1914/1916 to the 1917/1920 seasons by 10.7 % (Tables 16.3 and 16.4) and revenue by 23.5 % (Table 16.4). Revenue varied according to variety, quality and the proportion of each State's crop which was exported. The absence of production data for the other

Table 16.3 Apple production 1909–1920

State	Period	Orchard (acres)	Yield (bushels)	Revenue (£)	£ per Acre	£ per Ton
Q'ld	1914/1916	918	38,786	£15,277	£16.65	£20.94
	1917/1920	1,526	59,165	£34,955	£22.92	£31.52
NSW	1914/1916	6,502	501,326	£126,740	£18.45	£13.48
	1917/1920	7,836	726,089	£173,218	£22.10	£12.72
Vic	1914/1916	16,854	1,155,106	£318,543	£18.90	£14.71
	1917/1920	17,080	1,379,332	£321,992	£18.85	£12.49
	1909/1910	15,323	931,822	£256,700	£16.75	£14.69
	1916	16,973	2,953,968	£430,787	£25.38	£7.78
Tas	1914/1916	17,912	1,815,341	£352,940	£21.94	£10.37
	1917/1920	21,911	1,816,166	£431,383	£19.69	£12.67
SA	1914/1916	7,110	502,685	£101,066	£14.20	£10.72
	1917/1920	8,546	336,018	£ 93,189	£10.90	£14.79
WA	1914/1916	6,073	364,743	£136,421	£22.42	£20.06
	1917/1920	7,505	530,413	£243,166	£30.40	£24.45
All states	1914/1916	**55,368**	**4,377,987**	**£1,050,937**	**£16.40**	**£12.81**
	1917/1920	**64,404**	**4,847,183**	**£1,297,903**	**£17.87**	**£14.28**

Data obtained from Commonwealth and Victorian Government Reports

Table 16.4 Changes in apple production and revenue generated between periods 1914/1917 and 1918/1920 (based on data in Table 16.3)

State	Acreage	Yield	Revenue	£ per Ton
Queensland	66.0 %	52.5 %	128.8 %	50.5 %
NSW	20.5 %	44.8 %	43.5 %	−7.6 %
Victoria	1.3 %	19.4 %	38.6 %	−14.9 %
Tasmania	22.3 %	0.05 %	21.5 %	12.2 %
SA	20.2 %	−33.2 %	−7.8 %	37.2 %
WA	23.6 %	45.7 %	49.2 %	31.8 %
Australia	**16.3 %**	**10.7 %**	**23.5 %e**	**11.5 %**

States prior to 1914 puts limits on the confidence in assessing the impact of the investigation on their production histories. In Queensland there was no problem with bitter pit and as most apples were imported, the increases in production and revenue (Table 16.3) probably had more to do with the outbreak of World War 1 and the need to supply its own market than anything else. Apple production in Queensland is confined largely to the Stanthorpe district of the tablelands where soils are of high fertility and adequate available water holding capacity, which possibly explains the general absence of bitter pit in that State.

The severe drought years between 1916 and 1920 may have substantially decreased production in South Australia during 1917–1920 the affect of which was offset by the 16 % increase in the area of apple orchards and a 37 % price increase (Table 16.3) which offset the 33 % drop in yield and resulted in only an 8 % drop in revenue. NSW, Tasmania and Western Australia appear to have responded in a similar way to Victoria in adapting orchard practices almost as soon as McAlpine set up experiments to test them, especially regarding pruning and manuring.

In 1891, the first year fresh apples were exported, 3,250 cases, the equivalent of 3,348 bushels or 63 tons were shipped to London and realised £815 at £12/18/9 per ton. The average annual export between 1909 and 1911 was 1,057,710 bushels [19,833 tons] at £26/13/- per ton. By 1915, however, the price per ton of apples had risen from £26/13/0 to £42 per ton, due in large part to better quality fruit resulting from pre- and post-harvest control of bitter pit. Post-harvest cool storage on orchards, in warehouses and on ships, greatly enhanced the quality of fruit when it arrived on markets. Prices on the domestic Australian markets however were seldom more than half the export price and often less in the larger Australian markets (Table 16.3).

When the average annual yields for the Commonwealth over the three seasons following the end of the investigation (Table 16.3) are compared with that provided by McAlpine as his baseline data taken as the average over the three season prior to the start of the investigation (Table 10.1), there was an increase of 26 %.

16.3 Qualitative Evidence – The National View

In his First Progress Report, McAlpine (1912) suggested that the cause of bitter pit was complex, and that while each component had to be trialed in isolation, the various components would interact in unpredictable ways. In view of today's understanding of the cause, his insight was deeply perceptive. McAlpine's mindful analysis of his observations of practices used in the best orchards over the preceding two decades and his discussions with orchard supervisors and inspectors such as Carmody, Quinn and Turner led to this conclusion. Prior to 1910, McAlpine was aware of practices, which, if combined and used widely throughout industry, might reduce the impact of bitter pit. Much of what he did between 1911 and 1916 was examine and test each of these 'best practices'. He needed to know if they did reduce the incidence of pit and what their overall effectiveness was. Without the benefit of modern analytical techniques available through the development of statistical methods, meaningful results from such widely spread experiments containing so many unpredictable, inherent, natural variables, could have been difficult to evaluate. Ronald Fisher[A1] was not then even appointed to Rothamsted and it was not until 1925 and 1935 respectively that statistical analysis and experimental design as applied today, began to be understood. McAlpine's investigations finished well before such analytical methods were available.

In order to be fair to both McAlpine and the Advisory Committee, it is necessary to accept that there were successes and failures on both sides. McAlpine's main focus was on reducing the problem, but he also carried out various studies which he believed might help him gain a fuller understanding of the cause of the disease. Some of these efforts were based on inadequate procedures. For example, why did he set up an elaborate experiment to study the transpiration rate of the developing apple on two varieties *Autumn Pearmaine*, and *James Grieve* rather than *Cleopatra* and *Yates*, which he had used extensively for other comparisons of susceptible and resistant varieties? His findings were interesting, but did not advance understanding of the cause of bitter pit. On this basis, the Advisory Committee's recommendation that he include three susceptible and three resistant varieties in every comparison was sound advice that applied to this experiment as well as others. McAlpine orchestrated an extensive, multifaceted investigation spread over thousands of miles. The Advisory Committee did acknowledge his vast effort and the benefits of using cool storage and cool store transport, but they censured him for not identifying the cause.

16.4 The Prophet in His Own Land

In consequence of McAlpine's focus on lessening the problem, the advice contained in his 4th Progress Report (McAlpine 1915), orchardists, public servants and traders in horticulture and the fruit trades hailed the benefits of the advice McAlpine had put before them. Following publication of the 5th Report (McAlpine 1916), those expecting a revelation of the cause of bitter pit were disappointed and some criticism was made in the press. *The Hobart Mercury* (1916a, b) published two articles following the issue of McAlpine's Fourth Progress Report. One was clearly by a crank, but the other, a lengthy article, emphasised the work done in Tasmania which had provided quantitative evidence of the pit-promoting effect of heavy pruning, which McAlpine had been involved in. Most of the article dealt with McAlpine's comments about the possible cause of bitter pit and quoted his introductory comment that knowledge at that time did not allow dogmatism regarding the cause of bitter pit, but that the immediate cause was concentration of cell contents due to water deficiency leading to localised cell deaths adjacent to vein endings. He explained his hypothesis of how the balance between water availability, translocation and transpiration influenced cell responses and was in turn affected by weather. In terms of the importance of the balance between water uptake and transpiration he was correct, but in terms of events taking place in the protoplasm of affected cells he was not. Significantly, however, no one else had a better explanation at the time.

The main reason for general disappointment was expressed in *The Leader* newspaper on 17th March (1917), in a republished critique of McAlpine's Fifth Report, which appeared in *Nature*. Keeble (1916) had said he began reading McAlpine's report with great anticipation of learning the great mystery surrounding the cause of bitter pit. He acknowledged the thoroughness and assiduity of McAlpine's work, but failed to find the revelation of the mystery of the cause. He too of course was making his analysis in the vacuum of knowledge of the importance of calcium availability and the many factors that influence it. The article in *The Leader* emphasised the view that in presenting his Fifth Progress Report McAlpine should have made it clear at the outset that he had not found the cause, but only established factors he believed contributed to it, in which case readers would have focused solely on valuable information about managing the problem. In hindsight, McAlpine may have agreed that this could have strengthened focus on his improvements to orchard management, fruit storage and transport. On the other hand, had he done so, his detractors on and outside the Advisory Committee would have regarded it as an admission of failure.

Following their review of Australian research into bitter pit, Carne et al. (1929) noted that McAlpine had identified two key components of the means of reducing the incidence of bitter pit. These were the importance of avoiding heavy pruning, and the importance of placing fruit into cool stores at constant temperature as soon as possible after picking. They considered McAlpine's work to be particularly notable. W.M. Carne was a Western Australian plant pathologist who had begun his career in Victoria, and was a close colleague of members of the Advisory Committee. Carne had worked with McAlpine in Victoria, was familiar with his fieldwork and was qualified to make a fair assessment of McAlpine's contribution.

In his final paper on the subject, McAlpine (1921) continued to stress the complex nature of bitter pit: (i) the importance of water balance in the developing fruit, and its influence on factors which promoted rapid vegetative growth [he referred to it as excessive sap flow], (ii) the range of susceptibilities among cultivars, and (iii) the different responses of cultivars grafted onto different rootstocks, as factors also affecting the incidence of bitter pit. It was to be many years before the significance of water stress and calcium deficiencies were linked in other disorders, such as blossom end rot of tomatoes (Hopkins et al. 2009), and also many more before science recognised the extent to which farming practices reduce the pH of agricultural soils, which in turn impacts on mineral uptake (Reeves et al. 1987).

Although he was reluctant to be drawn into it, McAlpine undertook the bitter pit investigation in a determined and methodical manner. At the outset of the investigation, in a letter to the Government Botanist, McAlpine (1911) outlined two proposed preliminary experiments. One was to compare the impact of pruning practices on the incidence of bitter pit, and the other to determine if some manurial treatments were causing problems. He also made the significant statement that at the end of the season fruit would be put into cool store for one month, because that was how long it took for bitter pit to develop in stores at ambient temperatures. In retrospect, and in the light of knowledge gained long after McAlpine retired, this observation was a pointer to part of the solution to the problem as the conditions required for the long-term storage of apples depend on carefully controlled storage temperatures and atmosphere, often with highly specific regimes for different varieties of apples. By 1915 Government cool stores in Victoria catered for 175,000 bushel cases, 105,000 case capacity being available at Victoria Docks, while private cool store facilities around the state catered for a further 500,000 cases.

16.5 International Comment

Overseas there was little but praise for McAlpine's recommendations and the benefits that flowed from them. He won great respect offshore from governments, educationalists and orchardists in South Africa, Great Britain, Europe, Canada, the USA and Argentina. Mr. D.H. Ross (1915), the Canadian Trade Commissioner in Melbourne, wrote to say he had received numerous requests from lecturers in Canadian colleges and from Departments of Agriculture throughout the dominion for him to procure copies of McAlpine's reports, which he said were of great interest throughout Canada because of their importance to the apple growing industry. He asked for multiple copies of all McAlpine's reports. He later wrote for copies of the final report (Ross 1917). There were similar requests and comments from similar institutions in other countries. For example, in early 1916, L.R. Hassler (1916) wrote from the New York State College of Agriculture to obtain copies of McAlpine's reports and told him that his work was highly regarded by plant pathologists generally throughout the United States.

In March 1917, Professor Spencer Pickering FRS, Director of Woburn Experimental Fruit Farm, wrote telling McAlpine he regarded the conclusion from his bitter pit work to be of great importance (Pickering 1917). McAlpine relayed this to his friend, Professor Orme Masson at the University of Melbourne. In 1919, he was awarded Foreign Membership of the Royal Swedish Academy of Agriculture in recognition of his contribution to science (White 1986). Also in 1919, on 28th July, W.J. Morse, a plant pathologist at the University of Maine, Orono, USA, wrote thanking McAlpine for copies of his reports and telling him they were of great value. Then on 6th September his long-time friend, Professor H.H. Whetzel (1919) of Cornell University wrote to tell him that he had almost won a highly prestigious French prize for advances in science, but that he had lost it on some technicality to another nominee who had worked on Potato Leaf Roll Virus. One might wonder if the 'technicality' was his lack of a conferred degree! In his short History of Plant Pathology, Whetzel (1918) claimed that McAlpine and Marshall Ward[A1] (Fig. 16.1) were the two great British plant pathologists of the 'Millardetian Period'.

Seven years after the investigation was terminated and McAlpine had withdrawn from public life, Professor Richardson[A1], then Dean of the Faculty of Agriculture at the University of Melbourne, chaired the meeting of the Botanical and Agricultural Section of the Pan-Pacific Science Conference held in Melbourne in March 1923. On 23rd March he wrote to McAlpine (Fish 1976) telling him that a resolution

Fig. 16.1 Harry Marshall Ward (1854–1906). G.C. Ainsworth (1981) *Introduction to the History of Plant Pathology*, Fig. 20, p 58

Harry Marshall Ward (1854–1906).

proposed by Dr. T.M. Enfield, Director of the Cawthron Institute of New Zealand, and seconded by Dr. E.J. Butler[A1], Director of the Imperial Mycological Institute [forerunner of CMI], Kew, England, had been carried unanimously saying:

> This meeting places on record its feeling of regret that Mr. D. McAlpine has been unable to be present at the meeting of the Pan-Pacific Science Congress. It expresses its deep appreciation of the value of his contribution to plant pathology and trusts that he may long be spared to take an interest in the extension of the Science which owes so much to his pioneering investigations.

The reason for McAlpine's not attending the congress can only be surmised. According to Wedge (1984), his finances were precarious by that time, and he had also cut himself off from the Victorian science fraternity. It may be of significance that the motion of regret and appreciation was proposed and supported by influential overseas visitors. One wonders if it was a mild censure of Australian authorities over the manner of McAlpine's treatment.

Following McAlpine's death, copies of his 1898 portrait were distributed to various Institutions. A letter written in October 1934 was sent from IMI by Dr. E.J. Butler (1934) to Dr. D.B. Adam thanking him for sending the portrait, which he said would be hung prominently in the Institute premises at Ferry Lane, Kew. Later that year when McAlpine's daughter, Mrs. Constance MacDougall, and her husband were in England, Dr Butler invited them to visit IMI where he showed them the portrait hanging in the entrance hall (Wedge 1961). It was still on display in the entry hall when John Walker and I spent time there in 1971. Other institutions also received copies for which they sent letters of thanks.

References

Ainsworth GC (1981) Introduction to the history of plant pathology. Cambridge University Press, Cambridge, 315 pp

Bünemann G (1972) Annotated bibliography on bitter pit of apples, Bibliographical Series of the Technological University of Berlin. No. 2. Published by the University Library, Berlin, 170 pp. ISBN 3 7983 0106 9

Butler EJ (1934) Letter [2-10-'34] from the Director of the Imperial Mycological Institute to D.B. Adam thanking him for sending the portrait of McAlpine which would be hung prominently in the Institute premises at Ferry Lane, Kew

Carne WM, Pittman HA, Elliot HG (1929) Studies concerning the so-called bitter pit of apples in Australia – with special reference to the variety "Cleopatra", CSIR Bulletin No. 41. Government Printer, Melbourne, 101 pp

De Long WA (1936) Calcium and boron contents of the apple fruit as related to the incidence of blotchy cork. Plant Physiol 12:553–556

Fish S (1976) Daniel McAlpine – a pioneer plant pathologist in Australia. Aust Plant Pathol Newsl 5(1):11–14

Garman P, Mathis WT (1956) Studies of mineral balance as related to Baldwin spot in Connecticut, Connecticut Agricultural Experiment Station Bulletin No. 601. Connecticut Agricultural Experiment Station, New Haven, Connecticut

Hassler LR (1916) Letter to McAlpine [18-1-'16] requesting copies of McAlpine's Progress Reports on bitter pit

Hobart Mercury Newspaper (1916a) An Article [14-3-'16] summarised the 4th Progress Report

Hobart Mercury Newspaper (1916b) A second article [14-3-'16] criticised McAlpine's failure to devine the cause of bitter pit and his lack of advancing the means of controlling it much beyond what was known in 1911. It also reported a Mr. Norman's assertion of knowing the cause and reasons for failing to reveal it

Hopkins W, Norman G, Hüner PA (2009) Introduction to plant physiology. Willey, London

Keeble F (1916) Bitter pit. Nature 98:137–138

Leader Newspaper (1917) Article [23-3-'17] based on F. Keeble's article in Nature

McAlpine D (1910) Report of the vegetable pathologist 1907–10. Department of Agriculture Victoria, Government Printer, Melbourne, pp 44–61

McAlpine D (1911) He wrote to the Government Botanist [8-8-'11] outlining proposed pruning and manurial experiments

McAlpine D (1912) Bitter pit investigation. First progress report – the past history and present position of the bitter present position of the bitter bit question. Melbourne Government, 197pp

McAlpine D (1915) Bitter pit investigations. Fourth Progress Report – the experimental results in their relation to bitter pit and a general summary of the investigation. Government Printer, Melbourne, 178 pp, 70 Figures

McAlpine D (1916) Bitter pit investigations. Cause and control of bitter pit. Fifth Progress Report. Government Printer, Melbourne, 144 pp, 73 Figures

McAlpine D (1921) Bitter pit on Apples and Pezrs. The latest results in preservation measures. Phytopathology 11:366–370

Pickering, Prof. S (1917) Letter written to McAlpine expression his opinion of the great importance of the outcome of the bitter pit investigation

Reeves TG, Smith GH, White DH (1987) Integrated crop livestock production, pp 145–158 [See pp 149–150]. In: Connor DJ, Smith DF (eds) Agriculture in Victoria. Published by The Australian Institute of Agricultural Science, Melbourne, 231 pp. ISBN 0 85856 78 X

Ross BH (1915) Letter [13-12-'15] wrote to Dr. Cameron saying there was wide interest in McAlpine's investigation among Canadians and requesting multiple copies of McAlpine's Fourth Report

Ross BH (1917) Letter [13-4-'17] to Dr. Cameron requesting 18 copies of McAlpine's Fifth Report, and commenting in the wide interest in, and great value placed on McAlpine's reports within Canadian institutions

Sorauer P (1909) Handbuck der Pflanzenkrankheiten und Pflanzenschutzs [Handbook of plant diseases and plant protection], 3rd edn. Published by Eugen Ulmer, Stuttgart.

Thiem H (1936) Die Stippenkrankeit bei Lageräpfeln [Bitter Pit disease in stored apples]. Gartenbauawirtschaft 53:4

Warrington K (1934) Studies in the absorption of calcium from solutions with special reference to the presence of absence or boron. Ann Bot 48:743–776

Wedge E (1969) Notes held in the LaTrobe collection, signed by Wedge [31–5,'1961]

Wedge E (1984) Letters [4th to 7th May 1984] written to D.G. Parbery. Held in The LaTrobe collection

Whetzel HH (1918) An outline of the history of phytopathology, with 22 portraits. Published by W.B. Saunders Company, Philadelphia/London, 130 pp

Whetzel HH (1919) A letter [6-9-1919] to McAlpine, telling him he had come close to winning a prestigious Prize awarded for research. It was lost on a technicality to a French scientist for his study of Potato Leaf Roll

White NH (1986) McAlpine, Daniel (1849–1932). Australian Dictionary of Biography 10:193–194

Chapter 17
A Century Later

17.1 The Test of Time

After a century of investigation of the cause of bitter pit, there is no evidence of an infectious agent being involved, nor of the involvement of a toxic agent. The disease is a physiological disorder caused by several interacting factors, the mechanisms of which are still subject to investigation (Dr. Robert J. Holmes, pers. comm. 2011). There has been little added to knowledge about its cause since Snowdon (1990) reviewed the biology of bitter pit. She concluded that its cause is complex and is still being investigated, although broadly speaking, it is considered to be a mineral disorder resulting from an imbalance of essential minerals, there being two hypotheses as to how this comes about.

The first proposes that during normal plant metabolism, organic acids which form in the vicinity of vein endings are rendered harmless by the neutralizing reaction of calcium compounds distributed throughout the plant, including the fruit. If the supply of calcium is insufficient and conditions favour shoot growth, calcium is transferred from fruit to shoots. As the organic acids continue to accumulate at the ends of the vascular strands where there is now a calcium deficiency the acids are no longer neutralized, surrounding parenchyma cells die and turn brown. Calcium is an essential component of cell walls. Its balance with other minerals such as, magnesium and potassium affect its uptake and distribution, while abundant water and available nitrogen favour shoot growth.

The second hypothesis suggests that rapid redistribution of calcium to the core from peripheral tissue results in localised deficiency and consequent cell death. It was shown earlier that when there is active shoot growth in circumstances where calcium is in short supply, the actively growing shoot becomes the stronger sink for calcium, withdrawing it from the fruit. Besides the nutrient status of the soil, orchard practices which predispose fruit to bitter pit, include the promotion of excessive tree vigour and excessive fruit thinning. This is exacerbated by hot dry weather, especially if accompanied by wind. Apples are more prone to the disorder if immature

D.G. Parbery, *Daniel McAlpine and The Bitter Pit*,
DOI 10.1007/978-3-319-09552-3_17

when harvested. Also, large fruits from light crops are particularly susceptible All these conditions are conducive to low concentrations of calcium in the fruit. The rate of development of bitter pit after harvest is determined by storage conditions.

Neither of these hypotheses mentions the possibility of competing sinks for minerals and or other resources, between actively growing shoots and the developing fruit. At one point McAlpine suggested that there was such competition for water between fruit and shoot, and 20 years or so later, De Long (1936), noted that the only mineral difference between pitted and healthy fruit was the low calcium content of pitted fruit, compared to levels in healthy fruit. He concluded that the underlying reasons for this was competition for calcium between the growing shoots and young fruit, where the fruit failed to compete with the shoots.

Fifty years later, Snowden (1990) noted that attending to just one of the interacting factors resulting in the development of bitter pit did not prevent it because the cause is complex, echoing McAlpines view expressed 80 years earlier. Snowden also noted that soil analysis might indicate a need for applications of lime or other minerals. It may be advisable to modify pruning practices, and during fruit formation special care is needed to avoid water stress. Leaf and fruit analyses can assist in predicting susceptibility to bitter pit. It may be necessary to spray trees several times with a calcium compound and, immediately after harvest, to drench the fruit with calcium chloride. Only mature fruit should be picked, and delaying the harvest affords a measure of control. Storage conditions which delay fruit senescence are beneficial such as low temperature and controlled or modified atmospheres.

Since 1990, the understanding of the biochemistry behind the cause of bitter pit has changed little if at all. Control measures changed little over the following decade except that the need to dip fruit in a solution of $CaCl_2$ in as short a time as possible after picking is now emphasised and other refinements of the culture and post-harvest treatment of fruit is evident (Little and Holmes 2000). Since then, there has been little advance on this (Dr. Robert J. Holmes, pers. comm.).

Time, however, has vindicated McAlpine's approach to limiting the ravages of bitter pit. This is demonstrated by comparing his recommendations on the subject (McAlpine 1915, 1921) with those of today as set out by Snowdon (1990) (Table 17.1). It can be seen that McAlpine's recommendations have stood the test of time and are still valid.

17.2 Circling Around the Cause

Although Snowden (1990) nominated seven interaction factors that influenced the onset of bitter pit, McAlpine had recognized five of them. He had recognized the competition between actively growing shoots and developing fruit for water. Had knowledge of the physiology of senescence and ripening, and source-sink relationships been available to him, he may have got closer to understanding the cause of bitter pit, but neither of these areas of fruit biology had been studied. As a keen observer of natural phenomena over many years and possibly to some extent

Table 17.1 Control recommendation for bitter pit today and a century ago

Snowdon (1990)	McAlpine (1915)
Soil analysis for nutrient status to avoid high nitrogen levels and other imbalances.	Set up manurial trials, advised against over- use of nitrogen-rich manures.
Advisable to prune moderately.	Recommended light pruning.
Avoid water stress especially during fruit formation.	Recommended ensuring adequate water supply during the growing season, but stressed the need to avoid over watering.
If leaf and fruit analyses indicate a need, spray trees several times with CaCl2 or other recommended calcium compound.	The role of calcium was unknown at the time. McAlpine made no reference to it other than to raise soil pH.
Immediately after harvest drench fruit with CaCl2.	The role of calcium in fruit metabolism was 40 years and more ahead.
Harvest only fully mature fruit	Harvest only fully mature fruit
Delaying fruit senescence is beneficial, including temperature and controlled atmosphere.	Advocated immediate cool storage after picking. Controlled atmosphere had not been considered for fruit stores in 1916[a].

[a]By 1922 McAlpine was aware of the need to ventilate storages (see, Sect. 19.5)

intuitively, McAlpine knew that mature fruit was less likely to develop pit when stored at a constant low temperature. At the time, it was not known, in plant physiological terms, why this was the case. McAlpine also knew that different apple varieties had different keeping qualities and that good-keeping apples were the least susceptible to pitting. Cobb had observed this in the 1890s. Perhaps it was the converse! McAlpine had also observed that varieties of apple grafted onto different root-stocks often exhibited different rates of pitting. This suggested that the graft-unions in some combinations of scions and rootstocks were more efficient than others in the uptake and transport of minerals and water. Based on the information available he saw the selection of the best combinations of rootstocks and scions, and breeding, as the best future research prospects for the industry.

In his final paper, McAlpine (1921) placed great emphasis on the potential benefits of breeding pit-resistant varieties. He had observed substantial differences in the incidence of pitting, not only in different varieties, but also in the same varieties in different localities. He noted the absence of pitting in the wild ancestor of the apple, the crab apple. Thus his 5 years of investigation affirmed his initial view that bitter pit was constitutional rather than an infection or poisoning and that some varieties were more genetically predisposed to it than others. He also understood that the careful moderation of cultural and storage practices could reduce the incidence of the disorder.

When McAlpine's observations about the onset of bitter pit are considered within the context of today's understanding of plant physiology, it is evident that he had identified some key aspects of what was happening but was unable to connect them into a comprehensive explanation. This led to confusion, causing him to vacillate between thinking cells died because of water stress, or that they were killed when they burst under excessive sap-flow pressure. At one stage he did suggested that developing fruit was damaged when unable to compete for water and other resources against vigorous shoot growth. In the absence of the key findings that emerged 40–50 years later he was unable to draw his ideas and observation together and understand the process. He knew what it was not, while he sensed that in some way it was tied up with water availability, the balance between uptake and transpiration of water, flushes in growth (sap flow) and competition for resources between shoot growth and the developing fruit. However he lacked an intellectual framework on which to bring it together. In his penultimate report to the Advisory Committee he (McAlpine 1915) summarised what he had gleaned from the literature and what he had concluded from his experiments as follows:

1. There was disruption of the vascular network around the pits.
2. There were starch grains in pit-affected cells but not in adjacent healthy cells, as first noted by Cobb (1895), indicating a close connection between pit development and starch metabolism.
3. It was a physiological disorder rather than an infection or infestation. He called it a 'constitutional disease'.
4. The disorder began in the vicinity of the vascular network.
5. Super-abundance of sap may cause it. [He had expressed this elsewhere as vigorous vegetative growth out-competing the developing fruit for resources, which has proven correct].
6. In pits, the pulp cells have collapsed, the contents are concentrated, and adjacent cells injured.
7. The concentration [number] of injured cells increased when the supply of water was inadequate.
8. Over-watering could cause pit; for example, its incidence was high in very wet years such as 1910.
9. The outstanding fact was that pit was worse when there were excessive swings in sap-flow and transpiration [such as good rain after prolonged dry spells].
10. The condition is sensitive to the varying nature of climate and variety.

Although his critics castigated McAlpine for not having found the cause of bitter pit, it can be argued that he moved closer to it than anyone else at the time. Garman and Mathis (1956) found the first clue that was followed up, 40 years after McAlpine had finished. In the early stages of the investigation he examined the anatomy of pitted apples (McAlpine 1911a) in comparison with that of pitted pears (McAlpine 1911b) and quinces (McAlpine 1912) as well as that of healthy fruit, and he was in no doubt that the same processes led to pit in each case. His theories on the cause of the problem, however, never advanced beyond that of water imbalance and excessive sap flow, which caused the death of cells around vein endings. This was the nearest he got to explaining the cause.

In the light of present day knowledge, each of his explanations, although contradictory when considered separately, were describing different aspects of the cause, for which he saw no apparent link. McAlpine's thinking was circling the problem, and he came close to understanding the physiological mechanism of the link between the two paradoxical observations when he recognised that there was competition for resources between growing shoots and developing fruit. He focused on water, but was confused by the fact that both water stress and water abundance increased the incidence of bitter pit. The explanation of how this is possible emerged from two discoveries. One is that calcium is an essential nutrient required for plant cell integrity (Garman and Mathis 1956). The second is the recognition of source-sink physiology in plants in relation to resource distribution. Plants growing in an atmosphere containing radioactive carbon dioxide, transferred most photosynthate from the leaves, the source, into sinks such as the growing points and roots. Livny and Daly (1966) demonstrated how the competitive power of sinks could be altered. Over following decades it has been demonstrated that the distribution of other resources including water, amino acids and mineral nutrients can also be altered by changes in relative sink strength. Furthermore, it is commonly noted that the extremities of water stressed plants, such as leaf tips and upper margins of leaves are the first to show symptoms of water stress. In the case of bitter pit, with the onset of flowering and fruit set the young fruit begins to compete with the shoots for increasing proportions of photosynthate, water inorganic nutrients, etc. It is not the unavailability of calcium alone that causes bitter pit. Uptake and availability are influenced by the availability of, and balance with, other nutrients. Too much nitrogen promotes shoot growth, and fruit is starved of calcium as it does not compete strongly against the shoots. Excess magnesium and potassium results in these elements being taken up by fruit in preference to calcium (Snowdon 1990). Mineral nutrients are transported in solution, and are influenced by water supply, as well as edaphic and physical factors (Bünemann 1965).

McAlpine was interested in nutrient balances and did look at varying proportions of potassium, boron and magnesium in nutrient trials, but was unable to deduce much from them. The availability of today's experimental designs and statistical analyses may have enabled him to get closer to understanding the cause, which can be explained as follows. Under optimal conditions of water availability, shoot growth slows as fruit develops. Since water is essential for photosynthesis, and as photosynthesis is essential to maintain plant function, water stress slows growth and development and, as stress increases, water, minerals and other resources are withdrawn from the extremities of leaves and fruit in order to keep the photosynthetic 'powerhouse' and general metabolism operating. As water availability increases above the optimal balance for fruit development, it can trigger growth surges in shoots, which increases their competitiveness for resources. In the case of bitter pit in apples, when calcium availability is low, and resources including calcium are either not taken up in sufficient quantities or are withdrawn from developing fruit, insipient or immediate bitter pit results. McAlpine suspected that in trees subjected to water stress, shoots were more competitive for water than fruit, but what he did not know was that when there was low calcium availability, calcium is also withdrawn from the fruit and translocated to shoots. The main cause of the death of cells at the vein-endings in the fruit is calcium deficiency, not a lack of water. Calcium is an essential component of the 'glue' that binds cells together while allowing water and nutrients to pass from cell to cell. That substance is calcium pectate.

Towards the end of his career, Anton de Bary explored the role of pectolytic enzymes in plant disease (De Bary 1887; Ainsworth 1976), but it was not until the 1920s that a wider interest in the physiological nature of plant diseases began to appear. Carré and Haynes (1922) studied the chemical nature of pectin in apples, while Brown and Harvey (1927), Gaümann and Fischer (1929) and others turned to the study of the physiological nature of diseases, which has continued into the twenty-first century. Even so, such interest was focused on fungal and bacterial diseases of plants rather than physiological disorders which were seldom accepted as the domain of plant pathologists (Ainsworth 1981). Understanding of physiological disorders of plants and fruit needed the development of knowledge of the physiology of plant processes, including ripening and senescence. The roles of mineral nutrients including that of calcium, their interaction with each other and with the vagaries of the physical environment, all influence the outcomes. The cause of bitter pit is not unique as it has parallels with other physiological disorders such as blossom-end rot of tomatoes.

References

Ainsworth GC (1976) Introduction to the history of mycology. Cambridge University Press, Cambridge, 359 pp

Ainsworth GC (1981) Introduction to the history of plant pathology. Cambridge University Press, Cambridge, 315 pp

Brown W, Harvey CC (1927) Studies in the physiology of parasitism X. On the entrance of parasitic fungi into the host plant. Ann Bot Lond 41:643–662

Bünermann G (1965) Die Fruchtqualität beim Apfel in Abhängigkeit von der Nährstoffzufuhr II.Versuche zur der Kationen- und Stickstoff-versorgung und irher Qualitätsbeeinflussung. [Fruit quality in apple dependant on nutrient translocation II. Examination of the influence on quality of the availability of cations and nitrogen]. Gartenbauwissenschaft 30:3–44

Carré MH, Haynes D (1922) XII. The estimation of pectin as calcium pectate and the application of this method to the determination of the soluble pectin in apples. Biochem J 16(1):60–69

Cobb NA (1895) Bitter pit of the apple. Agric Gaz NSW 6:859–861

De Bary A (1887) Comparative morphology and biology of the fungi mycetozoa and bacteria. (Authorised English Translation). Clarendon Press, Oxford, 525 pp

De Long WA (1936) Calcium and boron contents of the apple fruit as related to the incidence of blotchy cork. Plant Physiol 12:553–556

Garman P, Mathis WT (1956) Studies of mineral balance as related to Baldwin spot in Connecticut, Connecticut Agricultural Experiment Station Bulletin No. 601. Connecticut Agricultural Experiment Station, New Haven, Connecticut

Gaumann EA, Fischer E (1929) Biologie der pflanzenbewohnenden Pilze [Biology of plant-colonising fungi]. G. Fischer, Jena

Little CR, Holmes RJ (2000) Storage technology for apples and pears – a guide to production. Published by the Department of Natural Resources and Environment, Knoxfield, Melbourne. ISBN 0 7311 4466 X

Livny A, Daly JM (1966) Translocation in healthy and rust-affected beans. Phytopathology 56:170–175

McAlpine D (1911a) The fibro-vascular system of the apple (pome) and its functions. Proc Linn Soc NSW 36:613–625, 4 Plates, 21 Figures

McAlpine D (1911b) The fibro-vascular system of the pear (pome). Proc Linn Soc NSW 36:656–663, 3 Plates

McAlpine D (1912) The fibro-vascular system of the quince fruit compared with that of the apple and the pear (pome). Proc Linn Soc NSW 37:689–697

McAlpine D (1915) Bitter pit investigations. Fourth Progress Report – the experimental results in their relation to bitter pit and a general summary of the investigation. Government Printer, Melbourne, 178 pp

McAlpine D (1921) Bitter pit on apples and pears. The latest results in preservation measures. Phytopathology 11:366–370

Snowdon AL (1990) A colour atlas of post harvest diseases and disorders of fruits and vegetables, volume 1 – General introduction and fruits. Wolfe Scientific Ltd, Barcelona, 302 pp. ISBN 7234 0931 5

Part III
Twilight of a Career

Although McAlpine stated that he had taken the investigation as far as possible at the end of 1915 and again in 1916, by the beginning of 1917 he was evidently keen to continue with aspects of the grafting and crossbreeding experiments established in 1912, which by 1917 were only just beginning to yield results. It seems that because the Prime Minister had stopped the investigation at the end of 1916, the Advisory Committee considered it had no further responsibility regarding McAlpine, and as he had been required to resign his post with the Victorian Department of Agriculture in 1911, it also took no further responsibility for him. Thus he was left in a state of isolation, confusion and anxiety as he tried almost desperately to find support to continue.

From early in the investigation he suspected that at lease to some extent, the root stock onto which apples were grafted influenced their susceptibility to bitter pit and it was only at the end of 1916 that his experimental grafted trees had flowered. Similarly the progeny of his cross breeding experiment between *Cleopatra* and *Yates* apples had just flowered. One can only imagine the utter frustration of not being given the opportunity of seeing the results of so much earlier effort.

His agreement with The Hon. George Graham, his Minister in 1911, forgotten and therefore not honoured, and his political master's deeming he had failed by not finding the cause of bitter pit, appear to have isolated him from further public service. In consequence he retired deeply hurt, disappointed and angry, and died more or less in obscurity. Possibly, the international delegates at the Pan-Pacific Science Congress held in Melbourne in March 1923 expressed their disapproval of McAlpine's treatment, when they moved a motion of appreciation of the man they so greatly admired.

Fortunately, his daughter and admirers such as Stan Fish kept the flame of his memory alive and eventually it was restored to full brilliance and kept burning by the membership of the Australasian Plant Pathology Society.

Chapter 18
Personal Interactions

18.1 Alfred Ewart

When Alfred James Ewart[A1] was appointed jointly as Professor of Botany at the University of Melbourne and Botanist to the Government of Victoria in 1906 he may have expected to be the main authority on matters botanical and the teaching of botany in the State. He was 34 years old, ambitious, impatient, short tempered and proud of his position. He could appear careless with the truth and did not always let facts stand in the way of a preferred theory. He was a highly qualified scientist with wide-ranging interests, well versed in taxonomy and plant physiology, including water uptake, transport and transpiration. He is perhaps best remembered for his compilation of *Flora of Victoria*. While being highly active and productive, he was not always a comfortable employer or colleague, especially during the early years of establishing himself (Cohn 2005). His impatience, and his awareness of his status, sometimes impacted badly on relationships with staff and others. In the early 1920s this resulted in censure of him by the University's Chancellor and its council (Blainey 1957). A few years later, however, following a period of poor health and overseas leave, his attitude to staff improved, and from then until his untimely death from a heart attack while supporting his Carlton football team, he enjoyed better relations with those with whom he dealt. At the time of his death he was generally admired and respected by staff and associates (Fig. 18.1). Chambers (1981) provides a detailed account of Ewarts educational background and his contributions to botany.

18.2 Ewart's Antagonism to McAlpine

Daniel McAlpine and Alfred Ewart were both members the Field Naturalists Club of Victoria of which McAlpine's protégé Charles Brittlebank was also a member. Club membership included numbers of public servants whose responsibilities involved plant identification and general biology, and interested members of the

D.G. Parbery, *Daniel McAlpine and The Bitter Pit*,
DOI 10.1007/978-3-319-09552-3_18

Fig. 18.1 Professor A.J. Ewart (1872–1937) (Courtesy of the Royal Botanic Gardens Library)

public including amateur taxonomists. Ewart joined the club on 12th March 1906, no doubt out of interest but also as a base from which he could control and coordinate the dissemination of botanical information. At the Annual General Meeting in June 1907 he was elected to the Club's Committee and from 1908 to 1910 he served terms as Vice-President and President of the Club (Cohn 2005). Thus he was well placed to control and protect his area of responsibility.

Unfortunately, within 2 years of taking up his appointments Ewart had become antagonistic towards McAlpine (Grasby 1911). Possibly he resented the fact that McAlpine, whom he saw as unqualified in terms of degrees, was so highly respected as a scientist and academic. Whatever the reason, in spite of his reputation as a lecturer and mentor of students, McAlpine was never invited into the School of Botany, and none of Ewart's staff was encouraged to engage with him. Although Ethel McLennan entered the University as an undergraduate in 1910, began teaching botany in 1914 and joined the staff of the Botany School in 1915, later teaching mycology and plant pathology (Gillbank 2008), she claimed never to have met McAlpine (Parbery 1989). This is surprising, considering that they were both in the group photograph of the 35 delegates present at the meeting of the *British Association for the Advancement of Science* held at Melbourne University from 28th July to 31st August 1914 (Fig. 14.1). It is all the more surprising considering Professor Ewart and Charles Brittlebank, each of whom knew both McAlpine and McLennan, were also present (Parbery 1989).

Ewart's attitude towards McAlpine is evident from the written record. As mentioned earlier, in a letter written on 23rd May 1911, W. Catton Grasby (1911) reminded McAlpine of a comment he made about Ewart's antagonism in 1908, which Grasby connected to the announcement of White's lead arsenate poisoning hypothesis in 1911.

Ewart's initial appointments were for 5 years during which, while he fulfilled his roles satisfactorily, his manner offended many, some in high places. Within the

Department of Agriculture, Ewart reported to the Minister of Agriculture rather than to the Director of Agriculture, Thomas Cherry. This was an unsatisfactory arrangement and led to serious clashes between Ewart and Cherry (Cohn 2005). Cherry's contract was not extended beyond 1910, after which he was appointed to the Chair of Agriculture in the University of Melbourne. During 1909, Ewart had been led to understand that his appointment as Government Botanist would be renewed for a further 5 years from 1911, but during May 1911, he was shocked when the Acting Premier, W.A. Watt advised him that his appointment as Government Botanist would cease from 30th June 1911 (Cohn 2005). The apparent change in the Governments attitude to Ewart followed correspondence between Watt and Acting Prime Minister Hughes in which Watt cast serious doubt on the worth of Ewart's evidence for his 'poison' theory regarding the cause of bitter pit. Ewart's contract was not terminated, however and he continued under different arrangements until his University appointment became fulltime in 1921. Meanwhile, the University was also somewhat dubious about extending Ewart's contract beyond 1911 for reasons held in Council minutes (Cohn 2005). This related to the unacceptable manner in which he attempted to remove Isabel Cookson and Ethel McLennan from his staff and his affair with one of his female associates, not Dr. White (Cookson and McLennan, person information).

In a personal letter to the author, McAlpine's daughter, Mrs Eric Wedge (1984) confirmed Ewart's unpleasant antagonism towards her father. She wrote:

> At the time of my father's appointment in 1911, there was violent opposition from Professor Alfred J. Ewart of Melbourne University, [who] was annoyed and said that the appointment should have been given to a University man. Feeling ran so high that a four-man Board of Inquiry was set up. Ewart was always hostile [to my father].

While there is no reason to doubt that such an inquiry was held, no record of it has come to light. It may have been part of the reason that Ewart was warned his contract as Government Botanist would terminate at the end of 1911. An example of his snide methods of demeaning others is evident the in the Government Botanist's section of the Department of Agriculture Report for 1907 to 1910, where he deliberately demeaned McAlpine and other highly experienced, non-graduate Officers of the Department under the guise of thanking them. He stated:

> Owing to the great increase in the teaching required at the University as the result of the progress of the Agricultural and Veterinary Schools, it has not been possible to give any outside courses of lectures. Occasional single lectures have been given, however, at different times. I have to thank Messrs McAlpine, Pescott, and C. French, Jnr. for imparting much technical information to the agricultural and veterinary students during visits to their departments.

He was prepared to use McAlpine and other officers to cover lecturing commitments but referred only to their imparting 'technical information', rather than scientific information. He described McAlpine in this manner on other occasions. McAlpine and Pescott were pioneers in their fields as was Charles French Snr., each highly regarded as scientists by their peers and each delighted in the natural sciences, yet none of them held degrees. One might wonder if this was a deliberately demeaning description of their standing by an intellectual snob! Charles French Jnr., however, held a B.Sc. (Hons.) degree but was acknowledged in the same way.

As noted earlier, when the minister invited Ewart to suggest someone to lead the bitter pit investigation, he took a swipe at McAlpine for being no more than a 'technician' with narrow interests when he wrote: 'For this investigation, the services of an ordinary technical expert are not likely to be of much service. An obscure disease of this kind will need a broader view and deeper insight for its solution than can be expected from the expert limited to one small branch of work'. There is no doubt that McAlpine devoted much of his time to developing control programs and finding the technology to support them, either through his own efforts, from literature, or from his wide network of contacts, but to imply that he had a shallow knowledge of the science behind his efforts, in terms of what was known at the time, was absurd. In the event, Ewart's 'broader view' was very narrow regarding the cause of bitter pit!

By mid-year, McAlpine was aware that Ewart was often biased, unaware of the facts or careless with the truth. In a memorandum to Dr Cameron, McAlpine (1911) referred to a letter sent by Ewart to The Hon. G. Graham MLA (Ewart 1911), in which he stated, 'bitter pit was first noticed by Cobb, our lead pathologist'. It was another attempt to weaken McAlpine's reputation in the mind of the reader. No fair assessment of these two highly worthy scientists would rank one above the other. Ewart's statement was also wrong in that bitter pit had been described by Crawford (1886) 9 years earlier than Cobb's 1895 publication. Cobb (1895) was the first to apply the name 'bitter pit' to the disorder. McAlpine (1900) was certainly concerned about bitter pit by the mid to late 1890s and was conducting experiments on bitter pit by 1902 (McAlpine 1904). Ewart's comparison was probably made in part on the basis that Cobb had a PhD while McAlpine had no degree.

On 4th August 1911, only 3 days after McAlpine was formally appointed to lead the investigation into bitter pit, Ewart (1910) wrote as Government Botanist to Minister Graham forcefully reasserting his view that bitter pit was caused by poison, in the form of spray residues absorbed through the skin of developing fruit or from soil. Appended to his letter was a 1,300-word statement setting out the reason for his retaining this view. In the letter, Ewart submitted a request to publish the statement (Appendix 2). On 14th August, Dr Cameron sent the following response to Ewart.

> The Hon. the Minister directed me to acknowledge receipt of your communication of 4th inst. making a request to publish a statement on the recent controversy regarding Bitter Pit, and to say that, after perusing the statement, he is of the opinion that no good purpose would be served by its publication, which he therefore refuses to sanction.

Ewart had earlier requested the government chemist E.S. Scott, to test affected fruit for spray residues, and also on the 14th August 1911, Scott advised him that no traces of any pesticides were detected in affected fruit. He also advised Ewart strongly not to release his views to the public. Ewart responded by suggesting that the techniques used were not sensitive enough to detect damaging levels of residues. This information was included in the statement which he sought permission to publish. While he heeded the minister's advice, and did not submit his paper to a newspaper, he persisted with his 'poison' hypothesis and continued to push it within scientific circles. Ewart published five papers (Ewart 1913a, 1914a, b, 1915, 1917) on poisons and bitter pit throughout the bitter pit investigation, four with the same title, 'bitter pit and the sensitivities of apples to poisons'. In view of the lack of evidence to support the poison hypothesis, this behaviour was irrational and obsessive.

It is evident in Sect. 11.1 that Ewart was party to confidential aspects of the business of the Advisory Committee through his influence over Drs Norris and Cherry. It is unlikely that Ewart was unable to differentiate between the notional sketching of the main veins in Grew's drawing (Fig. 11.3), from the carefully detailed drawing and photographs published by McAlpine (Figs. 10.2, 10.3, 11.1, and 11.2). The main aspect of Grew's drawings is the concentrations of sclereide around the cores of pears and quinces and scattered through their flesh, so that it is evident Ewart's accusations of plagiarism against McAlpine were made knowing they were false, and that he deliberately misled Norris and Cherry, neither of whom were plant anatomists. When Cherry confronted McAlpine with a copy of Nehemiah Grew's book, showing an illustration he believed was vascular tissue of an apple, McAlpine pointed out that it was of a cluster of sclereide! McAlpine also advised Minister Graham that although Ewart had claimed there were no copies of Grew's *Anatomy of Plants* in Australia, there was one in the library at the University of Melbourne! In a final touch, typical of McAlpine's sense of humour, he told the Minister that if he would like to see Grew's book for himself, he was invited to visit McAlpine's home and view his private copy!

Ewart's attempt to discredit McAlpine in the eyes of the Council of the Linnean Society of New South Wales, advising them that Grew's publication predated McAlpine's finding by 230 years, did him no credit. The Council merely noted Ewart's letter and took no further action, stating that it did not wish to be involved in a 'barren controversy'. Members of the Linnaean Society Council knew McAlpine, and no doubt knew the difference between the anatomical components in contention! Also, Dr Maiden[A1] served on the Council and knew McAlpine. This episode raises questions about Ewart's state of mind at the time, but regardless of his motive, his behaviour created an unpleasant disharmony and lack of trust between McAlpine and senior members of the Advisory Committee.

Ewart often made wild assertions without knowing the facts, when it suited him. For example, during the Government's tests of his 'poison' theory, prior to the start of the bitter pit investigation, Ewart claimed McAlpine was unaware that the chemical tests in use were not sensitive enough to pick up the low concentrations of arsenic and lead involved. He also included several other elements not used in horticulture at the time. This was unlikely to have been ignorance on Ewart's part. Thus he ignored the expertise of the State's chemist and was possibly ignorant of McAlpine's standing as a chemist! When launching his poison theory, he claimed lead arsenate entered fruit through stomates in apple skin or was taken up by roots and deposited at the vein endings in fruit. Over the following 6 years he changed his story, not based on emerging scientific evidence, but moderated to counter awkward criticism. Ewart held to the poison theory throughout the investigation in spite of the fact that there was not a shred of evidence to support it.

Ewart was charismatic, outgoing, a devout football fan and a Carlton supporter. He was often impatient, short tempered, and asserted his position as the State's top botanist. He appears to have disapproved of the reputation McAlpine had gained during the 20 years prior to his own appointments. Ewart and Cherry shared a different social world to that of McAlpine. They were part of the University establishment, which McAlpine was not. Similarly, each chairman of the committee, Dr Norris and Dr Cumpston, were Directors of Quarantine within the Department of Trade, housed in Spring Street, Melbourne, and were from medical backgrounds. Ewart was more likely than McAlpine to have operated in the same social and political circles as Cherry and Norris. McAlpine focused on plant pathology, farmers' welfare and his family, more than on politics.

Throughout these difficult years, McAlpine never made a personal attack on Ewart, but he did strongly attack the poison theory. On 11th September 1912 he addressed the Australian Fruit Growers' Conference held at Somerville on the Mornington Peninsula in Victoria. A lengthy report in *The Age* included much of his address to the conference. This not only included a strong swipe at the poison theory but also provided a taste of 'McAlpine the Lecturer – Public Speaker':

> Mr. McAlpine prefaced his remarks by stating that the investigation was a 'tough job' and that while many people might imagine it to be monotonous work, he assured the conference that the further he advanced the more fascinating it became. He was not ready to submit a definite statement on the cause of bitter pit, but he was able to state that many misconceptions had been swept away. It was a fundamental principle in plant pathology to know the structure and function of the plant under investigation, and that when the normal is known; it is easier to understand the abnormal. Many investigators appear to be content with the knowledge that an apple possesses skin, 'flesh', and core, but with the aid of chemicals he had removed the 'flesh' from the 'bones' and there remained a most beautiful network of sap-distributing vessels. This network ball, Mr. McAlpine explained, is attached to spiral vessels, two of which are opposite each seed and through which the sap rises from direct passages in the centre of the stalk. Photographs were shown in which the structure of these channels was plainly outlined, and Mr. McAlpine had no difficulty in convincing his audience that, to use his own words, he held a key to a solution of the problem, but that the lock was not yet opened. He pointed out that the disease made its appearance under the skin and that the vascular tissue around which the 'flesh' of the apple is produced, conveyed the sap to the outer boundary of the fruit, or in other words, to where the most rapid growth was taking place. The fruit on young and vigorous trees is liable to develop bitter pit because it [the young tree] is well nourished and growing so rapidly that the mesh of the vascular network is broken, this probably causes the disease. 'I give no decision' Mr. McAlpine remarked, 'but I am stating facts and even if these structural investigations fail to furnish definite conclusions as regards the cause of bitter pit, the structural delicacy of the vascular system that has been disclosed is sufficient to justify a demand that apples should be handled with all the care that is taken when packing eggs.'

He then spoke about the susceptibility of various pome fruits to bitter pit and work he was doing in comparing highly susceptible *Cleopatra* apples with the comparatively immune *Yates* variety and experiments with pruning, manuring, green manure crops, irrigation and root stocks which would yield information in the longer term before addressing the poison theory.

> I can', said Mr. McAlpine, with pronounced emphasis 'make a definite statement in regard to the spraying theory. We have absolute proof that the disease is not due to insects, fungi or bacteria but growers throughout Australia are considerably perturbed when an announcement was published to the effect that Bitter Pit was due to the action of chemicals in the spraying material used by orchardists. I denied the statement, and when my opinion was challenged I determined to strike, and when I strike I do not use a mallet; I use a sledgehammer. Historically the spraying theory is destroyed. The late Frazer Crawford, of Adelaide recorded the disease as appearing in South Australian orchards as far back as 1886. This was before spraying was adopted in the orchards of the State. I have received a letter from Dr Galloway, Chief of the Plant Bureau in Washington, and he assures me that the disease was known in the United States long before spraying became a feature of orchard work. Tests in various orchards in Victoria and South Australia have furnished uncontriverted evidence of Bitter Pit in apples that grew on unsprayed trees, and further, I have treated apples with chemicals with the result that the skin showed marks exactly resembling Bitter Pit, but nevertheless it was not the disease that is causing so much loss to apple growers.'

McAlpine concluded his talk by telling growers of the success of the voyage of the *SS Somerset*, of the importance of understanding that an apple in store was still a living thing, which breathed and perspired and that the object of placing apples in cool store was to check respiration. He assured growers that bitter pit was an internal disease, due solely to internal causes. The conference delegates considered these announcements of the greatest importance. The Chairman of the Conference, Mr. Tully thanked McAlpine, and in moving a vote of thanks referred to the satisfaction growers felt in having the assurance that they need not fear the use of spraying materials. In spite of the overwhelming evidence against the poison theory, Ewart (1913b) came back with the argument that by lowering the storage temperature, the action of poisons was slowed down.

Dr Jean White made her position quite clear, after Professor Ewart announced their 'poison theory'. She let it be known that it was no more than a hypothesis and that she had not wished to make any public announcement about it until she had tested it for at least 12 months. After publishing the basis for the talk given to the Royal Society of Victoria by Ewart (White 1911), there is no evidence that she continued working with him on bitter pit. In 1912 she took a job in Queensland where her efforts eventually led to the first successful biological control program in Australia following her recommendation for the introduce of a parasitic moth to control prickly pear[A1].

It appears that White's response to the justified criticism levelled at Professor Ewart and herself for their haste in announcing an essentially untested hypothesis over-rode her loyalty to Professor Ewart and initiated an empathy for the steady approach taken by McAlpine. In September 1911 McAlpine had mounted a display on bitter pit at the Australian National Apple Exhibition, the purpose of which was not specified. It seems that it was a display of medium sized healthy fruit of a range of apple varieties set out to inform visitors which varieties were bitter pit prone and which were not. The exception was one large apple central to the display, because it was badly affected by bitter pit. It disappeared from the display during the exhibition.

McAlpine made extensive but fruitless inquiries in an attempt to retrieve it. When Jean White heard of his search, however, she told McAlpine that Professor Ewart had asked her to run some tests on a very large, bitter-pit-affected apple, of the same variety as taken from the display, a day or so after McAlpine's pitted display apple had disappeared!

18.3 McAlpine and the Advisory Committee

It is evident from the report of the Advisory Committee's second meeting held on 21st January 1914, that the committee was not entirely happy with the way the investigation was proceeding. In the 3rd point in the minutes it is stated: 'The Committee repeats it previous recommendation that the experiments now being conducted and the number of trees need to be as many as practicable'. Then under the 6th point it is stated that; 'Resolution 5 of the 1912 meeting is of sufficient importance to be re-affirmed; 'That the irrigation experiments should be rigidly controlled and that *inter alia* records embracing the following points should be kept''. Five parameters were then listed which they wanted measured or described, some of which called for meteorological data. In addition to his trials at Burnley, McAlpine had trials running on orchards in several apple growing areas of each of the six states, so there was much to supervise and large amounts of data to be collected and collated. Although the number of trees in some individual trials may have been small, the total number of trees in overall trials with the same objective was large. It is important to remember that Sir Ronald Fisher[A1] developed today's approach to designing experiments and the statistical treatment of results only between 1919 and 1925 to 1935. Until then, a conservative approach of including large numbers of individuals in experiments was the norm.

The bumbling attempt by Professor Ewart, Dr Norris and Professor Cherry to discredit McAlpine in respect of his elegant exposure of the vascular network of pome fruits must have raised doubt in McAlpine's mind as to the moral support he had from the Advisory Committee. Norris's demand for the time consuming monthly reports clearly irritated McAlpine, as did his not being kept informed of procedures required by Norris for having the reports published. It is probable that McAlpine had simply followed the usual procedure entrenched over 22 years, during which he had dealt with the Government Printer himself. McAlpine also resented not being kept informed of decisions and events that impacted on him. There is little doubt that events of 1912 put McAlpine and senior members of the Advisory Committee at loggerheads. McAlpine's monthly reports were little more than diaries of events with little in the way of outcomes, because from the beginning of the investigation, confidential information had been leaked to Professor Ewart either by Norris or Cherry. Although there is no evidence that Professor Ewart had direct influence on Chairman Dr Cumpston after Norris left, the Advisory Committee was vocal about what it considered negative aspects of McAlpine's work while remaining silent or saying little about important outcomes.

In the months before the meeting of the Committee in August 1915, in addition to submitting monthly reports, McAlpine wrote to the chairman keeping him abreast of what he was doing. On 8th March he wrote from Albany in Western Australia notifying his arrival there on 24th February from Tasmania, and that while the crops in Tasmania and WA were good, they were disappointing in the other states. He returned from Albany to Melbourne on 14th March, and in a later letter, apologised for not having his results available for a Committee meeting held on 16th March. He wrote to Cumpston again on 5th May 1915 telling him that he had been busy collating the results from the preceding year during visits to the various states. Then in July 1915, he was in Bathurst NSW examining experiments that were established in several orchards in that district. In 1911 and 1912, he had received letters from leading orchardists in many apple growing areas of the various states providing the names of several hundred growers prepared to participate in experiments and trials. Several of the letters contained the names of up to 30 or more orchardists. It is evident that McAlpine took advantage of many of these offers and that he made use of the communication channels provided by the local groups to engage in valuable extension work. His wonderful ability as lecturer combined with his experiences from research enabled him to present valuable information at conferences and meetings. This resulted in his being universally admired and his advice being greatly appreciated by the horticultural industry generally. Orchardists and the industry were deeply grateful to him for the information he provided, the improvements he made and the great benefits his investigation brought them. The irrefutable evidence of the value of McAlpine's efforts was the production of apples in the Commonwealth during 1916, which reached a record 6,591,952 cases, 51 % above the 1914 yield (Argus 1917). The Advisory Committee seems to have taken little notice of the impact of McAlpine's efforts which represented a 63.54 % increase in marketable fruit in Victoria during the 5 years of the investigation, with the increase for the Commonwealth being in the order of 43 %. It is probable that some of this increase was due to expansion of orchard areas during 1911 to 1914, and favourable seasons, but the outbreak of World War 1 and manpower shortages possibly curtailed it.

18.3.1 Divergent Objectives

McAlpine himself may not have realised just how close he had come to understanding the cause of the problem. Possibly sensing that the time required to determine the cause would weaken his chance of finding the best management practices for minimise the incidence of bitter pit, he concentrated his effort on the latter, evident from the introductory remarks to his first report. From the start, McAlpine's emphasis was on the second defined objective of the investigation, 'finding a cure'; while the Advisory Committee failed to see the progress it sought, possibly because its highest priority was focused on 'finding the cause' after which the means of control might be evident. McAlpine cut to the chase, possibly because he either felt he understood the cause or that its establishment would take more time than he could spare from his search for management solutions. He reiterated his approach in an article published in *The Argus* in April 1916 (McAlpine 1916), in which he addressed the question of how to increase production. He emphasised that one answer to this in regard to the fruit industry was to reduce wastage by minimising losses from various diseases. He then referred to the potential reduction in losses from bitter pit through the adoption of light pruning regimes (Table 15.2). His down-to-earth practical approach was the better one, since finding the cause at that time was essentially a futile hope because of the lack of knowledge of physiological disorders combined with limited knowledge of plant nutrition beyond the importance of nitrogen, phosphorus and potash.

18.3.2 Knowledge of Mineral Nutrients

In spite of this, had McAlpine been more focused on searching for the cause, he may have picked up clues that he overlooked. It has been said (Beverige 1957) that 'chance favours the prepared mind' and possibly this was the reason that McAlpine did not pick up on the possible significance of the benefit of dissolved bone in reducing bitter pit. He had already stated that he considered lime was valuable in assisting the breakdown of green manure crops and in raising the pH of acid soils. Ignorance of the importance of minor- and micronutrients and the importance of calcium as a plant nutrient, limited the chance that McAlpine, or anyone else at the time would find the cause of bitter pit. Because of this, the way was open for McAlpine's critics to influence, not only members of the Advisory Committee, but also others in high places who were not necessarily biologists but still in influential positions. Professor Ewart's continuous and often covert criticism of McAlpine appears to have gradually reduced the confidence of some of his political masters in McAlpine's ability as a scientist and the value of the practical advice he gave. It was not until the latter decades of the twentieth Century that a reasonable understanding of the cause, and comprehensive control of the condition, was achieved.

18.4 Daniel McAlpine and Charles Brittlebank

Charles Clifton Brittlebank was born in Derby, England, in 1862 and migrated to Melbourne in 1893 with his wife, widowed mother, his brother Thomas and Thomas' wife. Shortly after arriving in Victoria, Charles and Thomas established dairy farms at Myrniong, 40 miles from Melbourne towards Ballarat. Within a year or so, Mrs. Marion Day, a friend of Mrs. McAlpine's, who was Charles and Thomas Brittlebank's aunt, introduced the two families. Daniel and Charles became firm friends. They shared mutual interests in geology and natural history and there were regular visits between them. McAlpine frequently collected specimens such as *Uromycladium tepperianum* (Sacc.) McAlp. at Myrniong and established a one-acre experimental wheat plot on one of the Brittlebanks' farms. McAlpine fostered Brittlebank's interests and taught him microscopy and photography. Through their friendship Brittlebank joined the Field Naturalists' Club of Victoria, where Alfred Ewart, Charles French and Archibald Campbell among others were also members. Charles Brittlebank, a skilled water-colourist and illustrator, was soon engaged by Charles French to illustrate his six volumes on *Destructive Insects of Victoria* (1891–2013). A.J. Campbell also engaged him to illustrate *Nests and Eggs of Australian Birds* (1900). From about 1897, he also provided coloured illustrations for each of McAlpine's books ranging from *Fungus Diseases of Citrus Trees in Australia* (1899) to *Smut Diseases of Australia* (1910) and the progress reports on bitter pit (1913–1916).

From 1908, when he was appointed as McAlpine's assistant in place of Mr. Robinson, Brittlebank provided most of the photographs, photomicrographs and drawings of microscopic subjects for McAlpine's publications. After McAlpine took on the bitter pit investigation, Brittlebank continued to provide photographic assistance to him for the compilation of his progress reports (McAlpine 1915). Brittlebank gained his knowledge of mycology and plant pathology through working with McAlpine. In some respects Brittlebank was similar to McAlpine in being intelligent, energetic, not hesitating to grasp opportunities, and in having a broad interest in natural history. However, while McAlpine was non-aggressive, open and direct in his dealings with friends and colleagues and meticulous in his work, Brittlebank was not necessarily so open or thorough. In spite of his lack of formal tertiary education it is evident that Brittlebank was keen to have McAlpine's job, and on 20th April 1911, over 3 months before McAlpine's formal appointment to lead the bitter pit investigation, Brittlebank (1911) applied to Dr S.S. Cameron, for the position as Vegetable Pathologist for Victoria. Brittlebank's appointment as Vegetable Pathologist, however, was not Gazetted until 1913, a few months before McAlpine reached 65 years of age in January 1914, the age at which public servants retired. Whether or not this applied to consultants such as McAlpine is unknown, and may not have been taken into account (Fig. 18.2).

Erica Wedge (1984) alleged that Brittlebank applied for the job behind her father's back and that McAlpine did not discover this until the end of 1916. McAlpine expected to resume where he left off as Victorian Vegetable Pathologist in 1911, believing that Brittlebank was minding the fort until his return. It seems unlikely that McAlpine was not at least aware that Brittlebank was Acting Vegetable Pathologist during 1911 and 1912, or that he was confirmed in the position before the bitter pit investigation finished. Once McAlpine realised what had been done, and that Brittlebank had been appointed as a permanent public servant entitled to a Government pension on his retirement, both of which had been denied to him, he was angry and hurt. He cleared his Flinders Street office, collected his microscope and destroyed his manuscripts and personal papers. He was later given signed copies of Brittlebank's letter of application for his old job. McAlpine was deeply hurt by the discovery and it ended their friendship. McAlpine had no further contact with Brittlebank. He also cut himself off completely from the Pathology Branch. He was not farewelled in any way, which was in stark contrast to his departure from the Victorian College of Pharmacy where he was presented with a gold watch and his wife with a valuable vase, each highly prized by them.

Fig. 18.2 Charles Brittlebank at work, 422 Flinders Street (Courtesy of the Department of Environment and Primary Industries, Victoria)

Charles Brittlebank was a colourful character, recognised as an illustrator of biological subjects, an amateur geologist and an amateur pugilist as well as a plant pathologist. He was a competent plant pathologist but after McAlpine's departure he relied on Dr Ethel McLennan for the identification of lesser-known pathogenic fungi. Whereas Professor Ewart never invited McAlpine into the School of Botany at The University, he facilitated cooperation between Brittlebank and McLennan, an undergraduate in the Botany School from 1911 to 1914 and a member of his staff from 1915. From 1926, Brittlebank provided some lectures to undergraduates at the School of Agricultural Science (Parbery 1989) while McLennan provided lectures in both mycology and plant pathology (Gillbank 2008, 2010). Brittlebank provided McLennan with specimens for teaching purposes, and relied on her to identify fungi, particularly those associated with new diseases. With Professor Ewart's approval, Dr McLennan and Charles Brittlebank set up complementary libraries. Neither the University nor the Department of Agriculture had funds enough to establish adequate libraries for their purposes. Thus the pair built up a 'common' holding of books and journals, half at Flinders Street and half at Parkville. The legacy of this valuable cooperation lasted into the 1970s. One might wonder what other advantages might have resulted had Ewart embraced McAlpine in a similar manner and worked with him rather than against him, and whether the schism between McAlpine and Brittlebank would ever have occurred had he done so?

References

Argus Newspaper (1917) Article published 29th May 1917

Beverige WIB (1957) The art of scientific investigation, 3rd edn. William Heinemann Ltd., London, 178 pp

Blainey G (1957) A centenary history of the University of Melbourne. Melbourne University Press, Melbourne, 220 pp

Brittlebank CC (1911) Letter [20-4-1911] to Dr. S.S. Cameron. LaTrobe collection Folder 12772

Chambers TC (1981) Ewart, Alfred James (1872–1937). Aust Dict Biogr 8:448–450

Cobb NA (1895) Bitter pit of apple. Agricultural Gazette of NSW 6:859–861

Cohn HM (2005) Watch dog over the Herbarium: Alfred Ewart, Government Botanist 1906–1921. Hist Rec Aust Sci 16:139–167

Crawford FS (1886) Report on Fusiclacliums, Cocllin moth, and certain other fungus and insect pests attacking Apple and Pear trees in South Australia. Government Printer, Adelaide, 77 pp, p53

Ewart AJ (1910) Report of the Government Botanist. In: Sharp AT (ed) Report of the Department of Agriculture [Victoria] for the years 1907 to 1910. Government Printer, Melbourne, pp 28–37

Ewart AJ (1911) Letter written to Hon. G. Graham [4-8-‘11], VPRI McAlpine File

Ewart AJ (1913a) On bitter pit and the sensitivity of apples to poisons. Proc R Soc Vic ns 24:367–419

Ewart AJ (1913b) An article in the *Argus Newspaper* [9-1-13] reported a talk given at the Royal Society on 6th January in support of the Poison Theory of Bitter Pit in which Ewart claimed that the benefit of cool storage was because poisons act more slowly at cold temperatures

Ewart AJ (1914a) On bitter pit and the sensitivity of apples to poisons. Proc R Soc Vic ns 26:12–44

Ewart AJ (1914b) On bitter pit and the sensitivity of apples to poisons. Proc R Soc Vic ns 26:228–242

Ewart AJ (1915) On bitter pit and the sensitivity of apples to poisons. Proc R Soc Vic ns 27:342–349

Ewart AJ (1917) Cause of bitter pit. Proc R Soc Vic ns 30:15–20

Gillbank LR (2008) Fungi – mycology and plant pathology. Unpublished notes on the History of the School of Botany, University of Melbourne, 12 pp

Gillbank LR (2010) From system garden to scientific research: the University of Melbourne’s School of Botany under its first two professors (1906–1973). School of Botany, University of Melbourne, Parkville, 40 pp

Grasby WC (1911) Letter written to D. McAlpine [23-5-1911], VPRI McAlpine Documents File

McAlpine D (1900) Report for the year 1900. Department of Agriculture, Victoria. 36 pp, 11 Plates, 18 Figs., Government Printer, Melbourne

McAlpine D (1904) Correspondence relating to the examination of bitter-pitted fruit by Dr. Bull, Department of Microbiology, Melbourne University

McAlpine D (1911) Sent a Memorandum [12-6-‘11] to Dr. Cameron raising his concerns about the reliability of aspects of Professor Ewart’s presentation of the Poison Theory of bitter pit

McAlpine D (1915) Bitter pit investigations. Fourth progress report – the experimental results in their relation to bitter pit and a general summary of the investigation. Government Printer, Melbourne, 178 pp

McAlpine D (1916) An article published in the *Argus Newspaper* [20-4-‘16] and the *Age Newspaper* [22-4-‘16]

Parbery DG (1989) Ethel Irene McLennan 1891–1983: pioneer teacher of mycology and plant pathology in Victoria. Australas Plant Pathol 18:47–56

Wedge E (1984) Letters [4th to 7th May-1984] to D.G. Parbery, La Trobe Collection

White J (1911) Bitter pit of apples. Proc R Soc Vic ns 24(1):1–9

Chapter 19
The Fruitless Years

19.1 Termination and Withdrawal

At the end of 1916, McAlpine said he had gone as far as possible with the investigation and was looking forward to returning to the Victorian Department. At the same time, the Advisory Committee was not convinced McAlpine's conclusions were reliable and considered some needed further testing. They saw no reason to continue the investigation on the same basis, but suggested further work should be put into the hands of the State Departments, but with the Advisory Committee retained as a coordinating body. Up to this point, McAlpine (1912) was disappointed over the Advisory Committee's attitude to the outcome and their failure to recognise that he had confirmed the validity of the hypotheses he set out to investigate. He had provided industry with a set of recommendations which were judged of great value around the globe. Whether or not this, or the fact that the commercial yield of apples in 1916 was an all-time record for the Commonwealth, gave him satisfaction or not is unknown, but there is no doubt that he was relieved to be no longer under any obligation to the Advisory Committee. Sadly, his disappointment soon turned to deep hurt and anger when he discovered he was out of work, and his career was terminated.

19.2 Forgotten Gentleman's Agreement

It is possible, but rather curious that McAlpine did not know that Brittlebank's position was made permanent in 1913, otherwise he would have known that his old position was no longer available for him. It raises the question of why either Dr. Cameron or Brittlebank had not told him of the nature of Brittlebank's appointment. It could have been that because McAlpine was officially working under a Federal Government Contract from 1911 to 1916, the State Administration had taken him

D.G. Parbery, *Daniel McAlpine and The Bitter Pit*,
DOI 10.1007/978-3-319-09552-3_19

off their list, and that no one other than Minister Graham knew of the promise, which was so important to McAlpine. In that case it is likely the successive Ministers for Agriculture were unaware of Minister Graham's agreement with McAlpine. Minister Hutchinson[A1], replaced Minister Graham in late 1913, and possibly knew nothing of the agreement when Brittlebank's appointment was made permanent. Minister Hagelthorn[A1] succeeded Hutchinson 2 years later, holding the position from November 1915 to November 1917. He would have known that Brittlebank had been serving satisfactorily since 1911 and, in the unlikely event that he knew of the agreement, he may have decided not to consider the 67 year old McAlpine's request for any form of reinstatement. Hagelthorn's working connection with Professor Cherry (Graham 1983) may also have worked against McAlpine's interest. On the other hand, not long after his appointment as Minister for Agriculture, Hagelthorn had urged the Premier of Victoria to advise the Prime Minister to extent the investigation to the end of 1917.

19.3 Attempts to Continue Working – 1917

In spite of Hagelthorne's apparent support, however, in December 1916, Prime Minister Hughes (1916) informed the Premier of Victoria that on the advice of the Advisory Committee (Cumpston 1916), the investigation into bitter pit would terminate on 31st December 1916. He asked the Premier to advise him if the Victorian Government was prepared to agree to the Advisory Committee's recommendations (Cumpston 1916). The Government agreed to four of the six recommendations but not numbers 3 and 4 until it determined whether it would need to employ an additional officer to continue the experiments begun by McAlpine, and the Prime Minister was advised of this decision. It seems that McAlpine was not told of the Victorian Government's response to the Advisory Committee's recommendation, although he was aware that the investigation was to continue on a State-by-State basis, and he assumed he would have a role in it.

Having bought *Invercliffe* in 1915, McAlpine had invested much of his earnings in its purchase and development. Having always worked on contract, he had no pension. It is evident from correspondence between McAlpine and Dr Cameron prior to the start of the bitter pit investigation that McAlpine's preference was to continue as Victoria's Vegetable Pathologist, not as a consultant, but as a public servant, which would have afforded him financial security in his retirement. This was denied him, and although he had 4 years on the heady salary of £1,000 *per annum* and a fifth on £750, once he had fulfilled his contract he had no income and no security into his retirement other than what he had managed to save.

Despite his feeling of abandonment, McAlpine did not give up entirely on his hope to continue working in some capacity. Either late in 1916 or early 1917 he wrote to Mr. George Swinburne[A1], who had been Minister for Agriculture from

1905 to 1908, asking for his help in finding employment. Swinburne (1917) replied saying it was absurd that McAlpine should find himself unemployed, and suggested that the Bureau of Science and Industry might be able to help him. This had been established the previous year through the efforts of the Prime Minister Billy Hughes[A1]. A meeting of the Advisory Council of the Bureau was scheduled for c. 20th February to establish the Bureau as a permanent Commonwealth organization (Sydney Morning Herald 1917). The Bureau was the forerunner of CSIRO, established in 1926.

At the end of January 1917, McAlpine (1917a) wrote to Premier Peacock[A1] in a manner suggesting he believed he still had a role in the ongoing investigation of bitter pit at State level (Fig. 19.1).

> The Hon. Sir Alexander Peacock – Premier,
>
> Sir,
>
> I take the liberty of writing to you in connection with the Bitter Pit Investigation.
>
> As it is an interstate matter, the Prime Minister is dealing with it, but as there has been some delay in giving official sanction to the continuance of the Experiments and as the season is advancing you would expedite matters by informing him that there is a Grant on the Estimates of £125 as a contribution by the Commonwealth Government towards expenditure relating to [the] Bitter Pit Investigation. There is also a sum set apart for 'Bitter Pit' investigation in the last Supply Bill of the Federal Parliament and I see no reason for this delay.
>
> I propose to complete the Investigation this year and I enclose the last letter received from the Prime Minister's Department.
>
> I have the honour to be, Sir,
>
> Yours obediently,
> D. McAlpine

There are three surprising aspects to this letter. Firstly, why did McAlpine wish to continue working on bitter pit? Wedge's recall of his feeling of abandonment after discovering his old position was closed to him at the end of 1916 would suggest otherwise. Why did he not give up entirely rather than attempting to seek support to continue working on bitter pit? One possible reason was his somewhat precarious financial situation, and another was his interest in the potential benefits he hoped would flow from his experiments in cross breeding of apple varieties, and the wide range of rootstock/scion combinations he had grafted. These experiments were only just beginning to yield results.

The second surprising aspect is that at the conclusion of each of his 4th and 5th Progress Reports, McAlpine had stated he had taken the investigation as far as possible, and yet at the beginning of 1917 he was expressing his impatience to continue his involvement in the grafting experiments at Burnley Gardens. The last sentence in his letter implies that he was in no doubt that he was to continue for another year, based on the recommendation of Hagelthorn at the end of 1915. It is possible, however, that he mistakenly expected to be kept on by the Victorian Department as the officer responsible for continuing the investigation under the terms of the 3rd and 4th recommendations made by the Advisory Committee in its final report

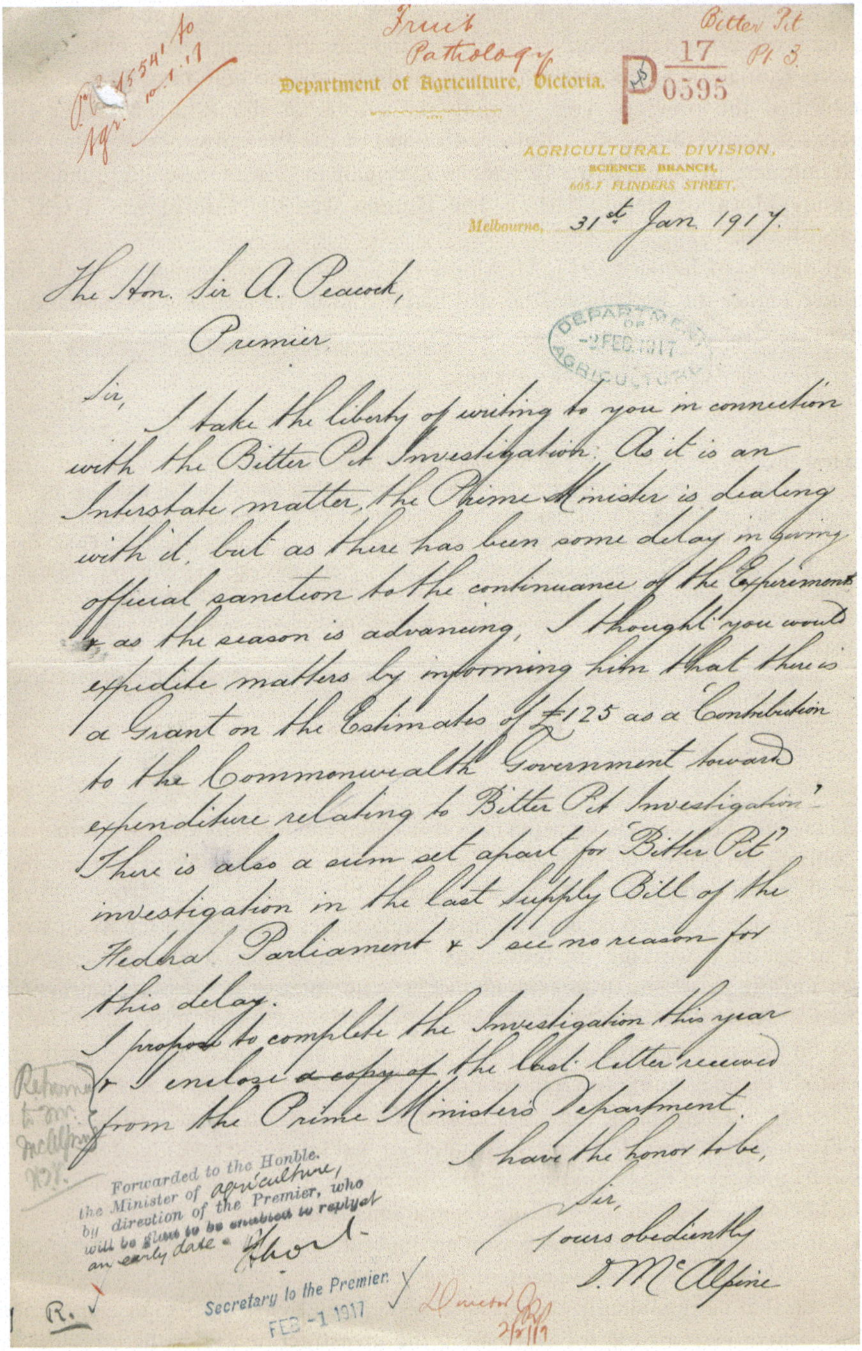

Fruit Pathology

Department of Agriculture, Victoria.

Bitter Pit Pt 3.

P 17 0595

AGRICULTURAL DIVISION,
SCIENCE BRANCH,
605-7 FLINDERS STREET,

Melbourne, 31st Jan 1917.

The Hon. Sir A. Peacock,
Premier.

DEPARTMENT OF AGRICULTURE -2 FEB 1917

Sir,

I take the liberty of writing to you in connection with the Bitter Pit Investigation. As it is an Interstate matter, the Prime Minister is dealing with it, but as there has been some delay in giving official sanction to the continuance of the Experiments & as the season is advancing, I thought you would expedite matters by informing him that there is a Grant on the Estimates of £125 as a "Contribution to the Commonwealth Government towards expenditure relating to Bitter Pit Investigation". There is also a sum set apart for "Bitter Pit" investigation in the last Supply Bill of the Federal Parliament & I see no reason for this delay.

I propose to complete the Investigation this year & I enclose ~~a copy of~~ the last letter received from the Prime Minister's Department.

I have the honor to be,
Sir,
Yours obediently
D. McAlpine

Returned to Mr. McAlpine

Forwarded to the Honble. the Minister of Agriculture, by direction of the Premier, who will be glad to be enabled to reply at an early date.

Secretary to the Premier.
FEB -1 1917

R.

Fig. 19.1 McAlpine's appeal to Premier Peacock (Permission of the State Library of Victoria, Melbourne – La Trobe Australian Manuscripts Collection [MS 12772, Box 3555/3])

(Cumpston 1916). His expectation of continuing under the new arrangements presupposes he was aware of them, in which case his expectation of continuing on the State's pay role is credible. Perhaps he had expected this throughout the investigation, because prior to its commencement, he had stated that some of the planned experiments could not be completed in the 4–5 years proposed for the investigation. Evidently he was not aware that the Victorian Government had agreed to continue work on bitter pit only if it could be done without appointing an additional officer.

Thirdly, the Prime Minister had advised the Victorian Premier on 13th December, that the investigation would be discontinued from 31st of December 1916. It appears to have been negligent of both Federal and State Governments not to have informed McAlpine that the Federal funding was solely to support the Advisory Committee as the central point for data collection from State-based on-going investigations, and collating them into an annual report to the Federal and State Governments. This apparent negligence may have been the result of the somewhat cumbersome administrative structure. There had been previous failures to inform him about decisions, which involved him, often discourteous to say the least. His request to Mr. George Swinburne for help in finding employment suggests a paradox in McAlpine's understanding of his situation. His letter to Premier Peacock confirms his expectation of continuing the investigation of bitter pit into a sixth year, whereas the correspondence with Swinburne implies that he was aware his involvement in the bitter pit investigation, and with the Victorian Department of Agriculture were at an end. This could have been an indication of the enormous stress he experienced by the end of the investigation.

A month after writing to the Premier (McAlpine 1917a) there had been no response. He consequently approached Dr Cameron, because on 5th February 1917, Cameron (1917a) sent a Memo to the Secretary to the Premier stating that Mr McAlpine should be informed of the contents of the Prime Minister's letter of the 13th December, and that the Victorian Government had agreed to discontinue the investigation. The following day, the Secretary responded in a postscript to Cameron's Memo to the effect that McAlpine had been informed, but did not state when or how. The reason for the confusion and administrative breakdown at the end of the investigation is unclear. Correspondence relating to it suggests that McAlpine was under considerable stress and was possibly suffering from mental fatigue and confusion. It is also clear that neither the Commonwealth Officers responsible for administering the bitter pit investigation nor State Officers took responsibility for advising McAlpine that the investigation was terminated. Either that, or none of them had the intestinal fortitude to face up to their responsibilities. In the end it was left to Dr. Cameron to clarify the situation. In the meantime, McAlpine was in a state of confusion and concern about his future.

On 16th March 1917, McAlpine (1917b) followed Swinburne's advice and wrote to his friend Professor Masson requesting his support to obtain a grant from the Bureau of Science and Industry. His letter is reproduced here as it explains some of the urgency behind McAlpine's appeal and his wish to ensure the collection of results from what he considered critically important experiments. They had run for 4 years and were on the threshold of yielding valuable results.

Dear Professor Masson,

I called to see you today at the University but unfortunately missed you. My Fifth Report on Bitter Pit is now issued and I will be sending you a copy in a few days. I have not only determined the cause of bitter pit, but devised a remedy and so far it is satisfactory. But from the nature of the case there are several important points, which cannot be settled in the time in an experimental way. There is for example the influence of the stock on the development of bitter pit, a question upon which all growers are interested, but it is only now that the trees planted at the beginning of the investigation are coming into bearing. The proportion of Pit in the produce of the different Stocks is the important point to settle and that can only be determined in the coming season.

I have a very fine set of experiments going on in a bird-proof cage at the Burnley Horticultural Gardens, but the Commonwealth Government does not intend to carry on the investigation further, as it seems to leave all these matters now to the Bureau of Science and Industry.

These are the only experiments connected with Horticulture being carried on in the Commonwealth and Spencer Pickering FRS [Pickering 1917], Director of the Woburn Experiment Fruit Farm writes to me concerning their great importance.

There are several experiments now underway and I was thinking the Bureau might be willing to make a Grant for their continuance.

Trusting to hear from you at your earliest convenience, as the experiments are now at a critical stage.

I remain,

Yours faithfully,
D. McAlpine

Professor Masson[A1] and George Swinburne had been early advocates of the establishment of an Advisory Council on Science and Industry. Professor Masson had been closely involved in its formation, and when McAlpine wrote to him, Masson was Deputy Chairman of the Advisory Council's Executive Committee. The Bureau had been established the previous year through the efforts of the Prime Minister Billy Hughes, and a few days after McAlpine wrote to Masson, the Executive Committee was due to meet in Melbourne (Sydney Morning Herald 1917) in order to establish the Bureau as a permanent Commonwealth organisation.

Professor Masson presented McAlpine's letter to the Executive Committee for consideration, but before making a final decision, the Committee requested advice from Dr Cameron (Lightfoot 1917a) regarding the final recommendations of the Commonwealth Advisory Committee on Bitter Pit. Lightfoot noted McAlpine's claim that the matter was urgent as the experimental trees were coming into bearing. Cameron (1917b) provided a summary of McAlpine's involvement in the bitter pit investigation from 1911 to 1916, and then in the fourth and fifth paragraphs he essentially cut the base from beneath McAlpine's chances by stating the Victorian Government's decisions on its future involvement in investigating bitter pit. Cameron informed Lightfoot that in December 1916, the Victorian Government had told the Commonwealth it would not be bound by the Advisory Committee on Bitter Pit's recommendations (3) and (4), until the financial obligations on the State and the call upon Departmental Officers' time was known, and would not be employing additional staff.

In any case, the Victorian Department was prepared to record trial results so long as it did not require the appointment of additional officers, or require so much of existing officers' time, as to interfere with their main duties. Accordingly, the Science Branch of the Victorian Department had been instructed to record results, as far as possible, especially in relation to experiments on the influence of rootstocks on the incidence of bitter pit, set up by Mr. McAlpine at Burnley Gardens in 1913. Cameron concluded the letter by drawing attention to the clash of opinion over the cause of bitter pit between McAlpine and the Advisory Committee.

On the same day, Cameron (1917c) sent a Memo to Mr. Laidlaw, reminding him of the Committee's recommendation the previous October to make arrangements to carry on with as much of the of work on the rootstock experiments at Burnley as necessary. He also advised that as Mr. Pescott had been closely involved in the Burnley experiments, he should be made responsible for the observations and continued investigations, particularly with regard to rootstocks. The implication behind these two items of correspondence from Dr. Cameron is that McAlpine was being shut out from further involvement in the investigation, not necessarily by Cameron but by his political masters. Not surprisingly, on 5th April, Gerald Lightfoot (1917b) acknowledged receipt of Dr Cameron's answer and told him that the Executive Committee had decided against providing a grant to McAlpine.

McAlpine had not received an answer from Premier Peacock to his letter of 31st January, requesting the £125, however, an *Aide Memoire* dated 11th April 1917, and initialed by Cameron (1917d) the following day, was headed 'Interview with Mr. McAlpine, at which it was explained: 1. That the vote of £125, included in the estimates for 1917, was specifically payable to the Commonwealth Government only. 2. That the State action was governed by the decision of the Commonwealth to discontinue the investigation from 31st December 1916, made on the recommendation of the Advisory Committee, consisting of all States in the Commonwealth.'

It seems most likely that it was in April 1917, following this run of events, that the bombshell of no longer having a job exploded into reality and not earlier as recollected by Wedge (1984). It may well have been after this interview that McAlpine returned home and told his family, 'Nobody wants me'! Wedge (1969) described her father as being like a ship adrift during 1916 and 1917.

On 9th April 1917, *The Age* reported that Mr. McAlpine had abandoned his concentrated sap theory of the cause of bitter pit in favour of a build-up of water pressure in cells causing them to burst and die. A month later on 11th May *The Argus* reported an address by Ewart (1917) to the Royal Society of Victoria in which he rejected McAlpine's burst cell and vascular interruption theory and reaffirmed his own belief in the slow poisoning theory put forward by Dr White. While McAlpine's 'burst cell theory' was unlikely, his original belief that the disorder was connected with the withdrawal of water from the outer tissues of the fruit was realistic and is now known to be one of the factors contributing to bitter pit and disorders of other of fruit involving the unavailability of calcium, such as blossom-end rot of tomatoes. As clever as he was in many respects, Ewart's antagonism to McAlpine blinded him to the potential of thoroughly investigating McAlpine's hypothesis regarding water stress and the death of cells surrounding the ends of vascular bundles beneath apple skin. Instead he doggedly clung to the poison theory in spite of there being no tangible evidence supporting it.

19.4 *Invercliffe* and Retreat into Retirement

Today, a person of McAlpine's calibre would be offered an appointment as a Senior Research Fellow at a University or relevant institution. With Professor Ewart and Professor Cherry in the relevant Chairs at Melbourne University that was highly unlikely. In December 1918, however, Professor Whetzel (1918b) invited him to visit Cornell University, Ithaca, NY, possibly as a Senior Research Fellow. By then McAlpine, at almost 70 years of age, was financially stretched and needing peace and time to himself after the intense activity and constant stress of the previous 8 years. Having been one who was not averse to promoting himself in a low-key manner and who thoroughly enjoyed interacting with those sharing his interests, he had lost his drive and no longer sought publicity. Possibly with some regret he declined Whetzel's invitation (Wedge 1984).

It is difficult to know why McAlpine was isolated to the extent he seemed to be, especially in Victorian political and University circles. In spite of the considerable progress he made in reducing the incidence of bitter pit, he had not identified its cause, and continuous criticism by his detractors took its toll on him and sowed seeds of doubt in the minds of key supporters in Government. Professor Ewart and Professor Cherry seem to have had enough influence within Government circles to make McAlpine's failure to find the cause of bitter pit be regarded as overall failure. Even Professor Waterhouse (1934), Sydney University, wrote to a Victorian colleague questioning whether McAlpine had been qualified to conduct the investigation into bitter pit. The irony in this was that McAlpine might also have felt he was not adequately qualified and he did not want the job, but no one else was found to do it. Professor Ewart had offered his services but, while the Victorian Governments provided a scholarship to support Dr White, there is no evidence to suggest Ewart was considered for the appointment. This is not surprising because at the time, the Administrations of both the University of Melbourne and the Victorian Department of Agriculture were seriously considering terminating Ewart's contracts at the end of 1911 (Cohn 2005) (Fig. 19.2).

During 1911, McAlpine and family moved from 22 Armidale Street, Armidale, to *Merchiston* in Mathoura Rd, Toorak, within walking distance of Hawkesburn Railway Station. In 1915, they bought a 40-acre farm, *Invercliffe*, on Queens Road, Wandin, some distance from the nearest railway station at Mount Evelyn, to which his daughter Erica drove him in their horse and buggy. He took the train into the city, 25 miles [40 km] to the west, when necessary (Fig. 19.3).

Wedge described *Invercliffe* as a botanist's dream, created by her father. There was a wide variety of trees from many parts of the world, many most likely from Nobelius's Nursery at Emerald. Wedge noted that the property was more like a botanical garden than a farm. McAlpine established a wide range of variously coloured rhododendrons and hedges of azaleas. All the plantings were labelled, giving their Latin and common names and countries of origin, which were written on wooden plaques attached to iron posts. There were chestnut trees with daffodils planted among them. He picked the chestnuts for sale and also grew strawberries, passionfruit, gooseberries and fruit, also for sale. After 1916, the sale of farm

Fig. 19.2 McAlpine family on the steps of *Merchiston*, Toorak in 1911 (From *L* to *R*: *Back Row*, Daniel, Constance, Edith Copon (friend), Isabel; *Centre* Isabella; *Front* Veronica and Erica. Permission, State Library of Victoria, Melbourne, La Trobe Australian Manuscripts Collection [MS 12772, Box 3555/3])

Fig. 19.3 McAlpine family in front of *Invercliffe*, Wandin in 1915 (From *L* to *R*: *Back Row* Gladys Rees (friend), Erica and Isabel McAlpine, E.E. Pescott; *Front Row* Isabella, Daniel and Constance McAlpine. Permission, State Library of Victoria, Melbourne, La Trobe Australian Manuscripts Collection [MS 12772, Box 3555/3])

produce was his only source of income from which he sustained the household and paid a farmhand to do the heavy work. They kept a milk cow and grew vegetables for their own use. His heart, however, was not in farming. He kept up regular correspondence with Professor Whetzel (1918a, 1919a, b) for many years and intermittent contact with other mycologists and plant pathologists including Lloyd[A1].

19.5 The Espiranto Dictionary

To help fill his days, McAlpine took up the study of Esperanto and spent many hours in the State Public Library in Melbourne where, using his knowledge of Latin and Greek, he found the derivation of every word included in an Esperanto Dictionary before sending the manuscript to Inazo Nitobe[A1] who was the Under-Secretary-General of the League of Nations. Nitobe was a keen advocate of the use of an international language with no political tie to any nation for all the work of the League. Apart from two letters from Nitobie (1926) in 1926 concerning the Esperanto language and McAlpine's Dictionary, nothing further is known about it. It most probably fell victim to the French veto of the proposal to adopt Esperanto as the official language for the League of Nations (Nitobie 1926).

Invercliffe still stands, however by 1924, their daughters had all married and were living away with their own families. So they suggested that Daniel and Isabella should sell *Invercliffe* and move closer so as to spend more time with them and their grandchildren (Wedge 1969). Eventually they were settled in Leitchville where Daniel McAlpine died on 12th October 1932; he was buried in the Cohuna cemetery.

19.6 Applied Evolution

Apart from publishing his conclusions about bitter pit and his views on further research into the problem, in *Phytopathology* (McAlpine 1921), McAlpine published nothing else. At about the same time, however, he wrote another paper presumably based on ideas developed through his interest in the possibility of breeding bitter-pit-resistant apples from crab apples, which were immune to bitter pit. He sent the article to the *Century Company* in New York expecting to have it published, but it did not appear in print. The reason is not evident. McAlpine's ideas may have been ahead of their time, or may not have been sufficiently developed. The manuscript has not been found, having either been destroyed or filed away in some unknown location.

Towards the end of 1916, the Director of Kew Gardens sent McAlpine a collection of crab apple seedlings McAlpine, further evidence that he was planning a breeding program. His objective was to select edible varieties of apples immune to bitter pit (McAlpine 1916). Of the dozen or so species of the genus *Malus* known as crab apples, most are tart but at least one is sweet. The most likely approach McAlpine intended taking was to cross crab apples with favoured domestic apple varieties, which were bitter pit prone. He may also have contemplated crosses and selections from within the crab apple species grouping, which was at the time believed to be the source of the domestic apple. He had had a long association with William Farrer and assisted in his cross breeding programs with wheat, which lead to commercially successful rust-resistant wheats as well as bunt-resistant and flag smut-resistant wheats. Consequently, McAlpine was fully aware of the processes and potential outcomes, which he had already applied to crossing bitter pit resistant and susceptible domestic apple varieties (McAlpine 1913).

Had he continued as Victoria's Plant Pathologist after 1916, it is evident that McAlpine would have continued monitoring the progeny of the crosses between *Yates* and *Cleopatra* apples. Whether or not he visited Burnley Gardens after 1916 is not known, nor is it known whether contact was maintained with colleagues such as Carmody, Laidlaw, Pescott and Turner after 1917. By 1920 they would have begun to see the benefits or otherwise of McAlpine's crosses, but no records have been found. It is possible, and to be hoped, however, that one or other of his colleagues kept him informed. The idea of crossing the resistant variety *Yates* with the highly susceptible *Cleopatra* was mooted by 1912 (McAlpine 1912). In the spring of 1913, McAlpine cross-fertilised the two varieties and by January 1914 he had collected pips from the crosses and sown them, with some seed already germinated by early February (McAlpine 1913). Four months later, in June 1914, he planted out the first of the resultant seedlings at Burnley Gardens. His last reference to the crosses was in his report of 1916 where he noted that the first of these had blossomed in the spring (McAlpine 1916). Following the termination of the enquiry, Messrs Laidlaw and Pescott were directed to take responsibility for the ongoing management of the trial but no further reports have been found. It is likely that, prior to the events leading to his abrupt severance from the Department of Agriculture, McAlpine had expected to be well placed to watch over the ongoing progress of his crossbreeding experiment, but sadly it was not to be. As it was, he was clearly still enthusiastic enough about the possibilities of crossbreeding to commit his ideas to paper and send it off for publication.

On 8th January 1933, 3 months after McAlpine's death, Mr. Barry Benefield, employee of the Book Editorial Department of the *Century Company*, at 353 Fourth Avenue, New York, wrote to McAlpine (Benefield 1933a) asking what he should do with the manuscript entitled *Evolution and Its Practical Application*, which McAlpine had sent to them in 1920 but which they had not published. As the letter was address to McAlpine's last work address at Flinders Street, it found its way to Burnley where it was dealt with by Dr. D.B. Adam, at the Biology Branch, which had been established in the Horticultural Garden, Swan Street, Burnley in 1929. On 31st March, Adam (1933) wrote to Benefield asking that the manuscript be sent to the Victorian Department of Agriculture Library, to which Benefield responded in April that (i) he would need authority either from the author or his executor, and (ii) he would need prepayment of the postage of $US 1.22 [=AU£0/5/7 i.e. 56¢]. He wrote again in May reminding Adam that he needed the prepaid postage (Benefield 1933b).

At the end of May 1933, Erica Wedge wrote to Adam saying her mother had received his request to authorise return of the manuscript to Adam. She was then living in Ganmain NSW, with her daughter, Claudine Cameron. Wedge (1933) reminded Adam that, for some time before his death, her father's memory had been failing steadily and that during that period he had written copiously to correspondents unknown to the family. She told Adam that as her father was not quite normal at the time, it had always worried them that such letters should have come into the hands of strangers. She felt sure that Mrs. McAlpine would be glad to give the manuscript to the library, but under the circumstances she thought it best if Adam should read it first, and presumably decide if it should be kept. Some years later, Wedge (1969) noted that her father's memory began to fail about 2 years before his death. Isabella, Mrs. I.J., McAlpine duly signed the authorisation, which was witnessed by Mr. Cameron, the Manager of the Ganmain Branch of the Bank of New South Wales, and her son-in-law.

In June 1933, Mrs. Claudine Cameron also wrote to Adam from Ganmain, sending her mother's authorization (Cameron 1933). She told him her mother had been very ill and was unable to write but she had asked Mrs. Cameron to tell him that although she had always been conversant with her husband's work, none of the family had had any knowledge of the *Evolution* manuscript. She was therefore afraid that it had been written unknown to them when his faculties were beginning to fail and was anxious for Adam to go through the manuscript and if her fears were confirmed, to destroyed in justice to her husband's real work. Claudine Cameron added that if it were appropriate to place it in the Laboratory Library, her mother would be pleased.

To date the manuscript has not been found, so it is possible that Mrs. McAlpine's fears were realised and that Adam destroyed it. However, since Benefield stated that the manuscript was sent to New York in 1920, at the time when McAlpine published a perfectly rational paper on Bitter Pit in *Phytopathology*, it would seem unlikely that the *Evolution* manuscript, prepared a year earlier, was not also a rational paper and may still exist. Furthermore, in the November following her father's death, Wedge (1932) wrote to Professor Whetzel telling him that Daniel had died peacefully in his sleep and that over the previous 3 years his mental capacity had declined. This was about 9 years after he had written the paper on applied evolution and when it seems his mental capacity was intact. During 1933, Adam was appointed as head of plant pathology at the Waite Research Institute of the University of Adelaide where he took up the position in 1934 (Edgeloe 1984). It is therefore possible that he had not dealt with the manuscript before he went to Adelaide. Even so, had he felt it appropriate, he would most likely have returned it to the Burnely library! One wonders why it took Benefield 12 years to return the paper. Possibly McAlpine refused to pay the postage after his disappointment over yet another rejection! Possibly also, those in the New York office failed to appreciate its insight into the possibilities of tree breeding and put it aside where it gathered dust.

Furthermore, in 1922 McAlpine responded to growers who had appealed for advice on how to prevent the development of *brown heart* in consignments of apples during cool storage transport oversea. In some consignments, fruit developed extensive necrosis, beginning around the core and extending out into the flesh. McAlpine (1922) advised that the condition resulted from the build up of carbon dioxide in storages unopened during voyages of several weeks, and that the storage holds should be opened every 3–4 weeks long enough to allow adequate ventilation. This advice was sound and appears to have relieved the problem.

References

Adam D (1933) Adam replied [31-3-'33] on behalf of Mrs McAlpine to Benefield's letter to McAlpine [8-1-'33]

Anon. (1881) Letter [30-9-1881], on Science Form 1134, headed, *Science & Arts Department of the Committee of the Council of Education, South Kensington*, listing the names of those who attended a Special course of Instruction in Agricultural Chemistry for Teachers at Rothamsted and passed Subject XXIV – *Principles of Agriculture*

Benefield B (1933a) Wrote [8-1-'33] to McAlpine about the return of his manuscript *Evolution – its Practical Application*, from the Century Publishing Company, New York

Benefield B (1933b) Wrote [April-'33] to Adam for family authority plus postage, and again in May '33 reminding Adam he needed postage

Cameron SS (1917a) Memorandum [5-2-'17] set to Secretary to the Premier of Victoria advising that McAlpine should be informed of the content of the Prime Minister's letter of 13th December 1916

Cameron SS (1917b) Letter [2-4-'17] responding to G. Lightfoot regarding information sought by the Executive Committee of the Bureau of Science and Industry regarding McAlpine's request for support, made to Professor Masson

Cameron SS (1917c) Memorandum [2-4-'17] sent to Mr. Laidlaw directing him to organise continued work and observations of McAlpine's Burnley experiments and for him to involve Mr. Pescott

Cameron SS (1917d) *Aide Memoire* [11-4-'17] regarding interview with McAlpine

Cameron C (1933) Letter [8-6-'33] to Dr. D. Adam, returning a signed authority for a manuscript to be returned from New York [La Trobe Collection -VPRI box]

Cohn HM (2005) Watch dog over the Herbarium: Alfred Ewart, government botanist 1906–1921. Hist Rec Aust Sci 16:139–167

Cumpston JHL (1916) Fifth Report of the Advisory Committee [30-10-16] circulated to the Prime Minister of the Commonwealth and State Premiers

Edgeloe VA (1984) The Waite Agricultural Research Institute: the first fifty years 1924-1954. Glen Osmond, 208 pp

Ewart AJ (1917) Theories of "Bitter Pit" – Professor Ewart's opinion. Argus Newspaper, 11th May 1917, 2 pp

Graham JW (1983) Frederick William Hagelthorn (1864–1943). Aust Dict Biogr 9:155–156

Hagelthorn, Hon. FW (1915) He wrote [13-11-'15] to the Premier recommending that he advise the Prime Minister to extend the bitter pit investigation to 1917

Hughes, The Hon. W.M. Hughes (1916) Circular [13-12-'16] sent to Premier of Victoria regarding the discontinuation of the Bitter Pit Investigation

Lightfoot G (1917a) Letter [22-3-'17] sent from the Acting-Secretary of the Executive Committee of the Advisory Council of Science and Industry to the Director of the Victorian Department of Agriculture regarding McAlpine's application for a grant

Lightfoot G (1917b) Letter [5-4-'17] informing Dr. Cameron that McAlpine's request for a grant had been declined

McAlpine D (1912) Letter written to Hon. Geo. Graham [4-12-'12] objecting to Norris's requiring monthly reports of his progress on the grounds that it was not required in his contract and that the information contained was forwarded to Professor Ewart

McAlpine D (1913) Bitter pit investigations. Second progress report – the cause of bitter pit: its contributing factors, together with an investigation of susceptibility and immunity in apple varieties, Melbourne, Government Printer, 224 pp, bibliography 96 pp, and 61 plates & maps

McAlpine D (1914) Bitter pit investigations. Third progress report – the control of bitter pit in the growing fruit. Melbourne, Government Printer, 176 pp and 77 figures

McAlpine D (1916) Bitter pit investigations. Fifth progress report -the cause and control of bitter pit. Melbourne, Government Printer, 144 pp, 73 Figures

McAlpine D (1917a) Letter [31-1-'17] written to Premier of Victoria regarding funding for his Burnely Experiments

McAlpine D (1917b) Letter (16-3-'17) written to Professor Masson, refuting the Advisory Committee's conclusion that he had not found the cause of bitter pit. [VPRI Box]

McAlpine D (1921) Bitter pit on apples and pears. The latest results in preservation measures. Phytopathology 11:366–370

McAlpine D (1922) Prevention of loss in oversea apple shipments. Fruit World of Australia, October 2nd, pp 334–337

Nitobie I (1926) Letters [7-9-'26 & 30-10-'26] sent from the Under-Secretary-General to the League of Nations concerning the Esperanto language and McAlpine's Esperanto Dictionary

Swinburne G (1917) A letter written to McAlpine [11-1-'17]

Sydney Morning Herald Newspaper (1917) An article reporting the permanent establishment of the Advisory Council of the Bureau of Science and Industry, 17th February 1917

Waterhouse LW (1934) A letter written to a Victorian colleague [3-2-'34] raising the possibility that McAlpine had not been adequately qualified to lead the bitter pit investigation

Wedge E (1932) Letter sent [3-11-'32] to Professor Whetzel telling him of her father's peaceful death and final years. [La Trobe Australian Manuscript collection MS 12772, State Library of Victoria]

Wedge E (1933) Letter [30-5-'33] sent to Dr. D. Adam regarding authority to have manuscript returned from New York. [La Trobe Australian Manuscript collection MS 12772, State Library of Victoria]

Wedge E (1969) Notes In the La Trobe Australian Manuscript collection MS 12772, State Library of Victoria

Wedge E (1984) Letters [4th to 7th May-1984] to D. G. Parbery, [La Trobe Australian Manuscript collection, State Library of Victoria]

Whetzel HH (1918a) A letter [13-11-1918] to McAlpine, thanking him for his appreciative comments on Whetzel's little book on the *History of Plant Pathology*

Whetzel HH (1918b) A letter [13-12-1918] to McAlpine, inviting him to visit the USA and spend some time at Cornel University

Whetzel HH (1919a) A letter [28-2-1919] to McAlpine, thanking him for his review of *Manual of Fruit Diseases*

Whetzel HH (1919b) A letter [6-9-1919] to McAlpine, telling McAlpine that he had come close to winning a prestigious Prize awarded for research. It was lost on a technicality to a French scientist for his study of Potato Leaf Roll

Chapter 20
A Reputation Re-established

20.1 Rekindling Appreciation

By 1949 it seemed that the enormous contribution made by Daniel McAlpine to plant pathology was disappearing from Australia's collective memory. Willis (1949) drew attention to the lack of tributes from the Australian plant pathology fraternity following McAlpine's death, which he clearly regarded as disappointing to say the least. The only ones found had been those from the Victorian College of Pharmacy, detailed in Chap. 4 and one from graduates in agricultural science from Sydney University probably written by L.W. Waterhouse (Anon. 1932) who was a committee member of the Sydney University Graduates Association (SUAGA) at the time. In the meantime, however, two Australian plant pathologists, Stanislaus Fish and Charles McGee, respectively Government Biologists for the Victorian and New South Wales State Departments of Agriculture, had decided to rectify the situation. Erica Wedge had provided Fish with her father's correspondence and notes on his personal history, believing something should be done to mark the Centenary of McAlpine's birth. In June 1949, Fish (1949) wrote to McGee, sending him the enlarged photographic portrait of McAlpine taken in 1898 plus some biological notes. In his response, McGee (1949) proposed using the notes to prepare a brief entry in the *Plant Disease Recorder* and suggested that action be taken to have the year of McAlpine's birth commemorated by the issue of a commemorative postage stamp in the same way as had been done for William Farrer in 1945.

By 1949 the appreciation of McAlpine's contribution to Australian Plant Pathology had waned except among small pockets of admirers who continued to remind readers and contacts of the valuable contribution he had made. For example, in the late 1920s, Carne et al. (1929) drew attention to the value of McAlpine's recommendations for bitter pit control. They seem to have been the only Australians to do so at the time. Had there been regular communication between plant pathologists then, his contribution including the benefits of the bitter pit investigation might have remained stronger in the collective consciousness than it did. But exchanges

D.G. Parbery, *Daniel McAlpine and The Bitter Pit*,
DOI 10.1007/978-3-319-09552-3_20

between pathologists in the different states were a long way off. Most communication was confined to correspondence between individuals and limited to plant disease subjects of mutual concern of the day.

In tracing the events which eventually led to the formation of the Australasian Plant Pathology Society, Dr Gordon Purss (1994) noted the few occasions when biologists interested in plant diseases were brought together. The first such events were the Rust-in-Wheat Conferences held between 1890 and 1896 in various states, but which did not continue as regular meetings between plant pathologists because prior to 1909 there were only Cobb, Bailey and McAlpine. In 1910, McAlpine (1910) published his views on the need for a strong, well-resourced Federal Establishment to deal with nation-wide plant disease problems and have them linked to the State Establishments. Such a national structure may have provided stronger meetings of minds and helped avoid rivalries and territorial jealousies that were evident even into the early 1970s. Purss (1994) reported attempts to bring plant pathologists together to discuss items of mutual interest, the first in 1923 and a second in 1927. However, either their employers failed to support the endeavour, or plant pathologists were not interested in the idea, since only two interstate delegates attended the Sydney meeting in 1927.

20.2 The Australasian Plant Pathology Society

In May 1935 the Federal Government established the Australian Agricultural Council for the purpose of consultation between the Commonwealth and States on issues affecting primary industries and primary products. At a meeting in Sydney in 1949, it took action of which McAlpine would have approved and which, over the following 20 years steered thought toward the establishment of the Australian Plant Pathology Society. At a meeting in Sydney in early 1949 the Australian Agricultural Council directed the newly reconstituted CSIRO to organise a conference of plant pathologists. The conference was held in Melbourne from 23rd to 27th May. One immediate benefit was that delegates decided to produce an informal publication, *Plant Disease Recorder,* for the purpose of keeping Australia's plant pathologists in touch with records of new or other interesting plant diseases, visits of interest from overseas and general news. This was distributed to government departments and universities where plant pathologists were employed. Other benefits of the *Recorder* were that delegates discovered others with similar interests, and most importantly, plant pathologists found time to consider the structure of their discipline.

The Melbourne conference was followed by similar conferences at Hawkesbury College, NSW in 1955, the Waite Research Institute, Adelaide in 1961 and Toowoomba, Queensland in 1966. Following the Adelaide Conference the idea of forming a society and running its own conferences at closer intervals began to take root such that at the 1966 Conference in Toowoomba, Professor Noel Flentje presented an out of session talk entitled *Australian Plant Pathology Society – is one desirable*? Delegates agreed that such a society was desirable and established a Steering Committee to draft a constitution. The Steering Committee met in Melbourne in 1967. In 1969 the *Australian Plant Pathology Society* was launched. Within a decade it was renamed *Australasian Plant Pathology Society* to embrace New Zealander members and acknowledge a wider membership within the region and beyond. In 1972 the Society began producing its own publication, *Australian Plant Pathology Newsletter*, replacing *Plant Disease Recorder*. The journal is now entitled *Australasian Plant Pathology* and published by Springer.

It was unlikely to have been a coincidence that within 2 weeks of the Melbourne meeting in 1949, Stan Fish and Charles McGee began working toward having a commemorate stamp issued to mark the century of Daniel McAlpine's birth. After Fish sent him the portrait and the idea of a commemorative stamp, McGee told Fish that he had the support of his Director, Dr Clive Noble, who had outlined the procedure to follow and urged that it be discussed with the Victorian Director of Agriculture, Mr. Ron Mullett (Purss 1994). He also recommended that the move originate from the Victorian Department of Agriculture. He proposed obtaining support from the NSW *Linnaean Society* and the *Royal Societies* of both NSW and Victoria. Their efforts came to nothing since the Post Master General declined to issue the Commemorative Stamp. That having failed, Fish decided to prepare a biographical article about McAlpine at about the same time as the editor of the *Australian Journal of Pharmacy* (Sissons 1968) was engaged in a similar exercise (Fish 1966). From 1920 to 1962 Alfred Sissons was Dean of the Victorian College of Pharmacy who shared many interests and qualities with McAlpine. Fish (1970) describing McAlpine's part in the history of plant pathology in Australia, and later a biographical article (Fish 1976). In an effort to get as much information as possible, Fish contacted McAlpine's daughter, Erica Wedge who, since her father's death, had been collecting his letters and papers with the intention of placing them in archives. Most were placed in the Department of Agriculture Library at Burnley Gardens and later, when the Burnley Institute was closed down in 1993, the collection was moved to the LaTrobe Collection of the Public Library of Victoria, the exception being correspondence between William Farrer and Daniel McAlpine, which she placed in the Mitchell Library of NSW in Sydney.

As Erica Wedge was living at home with her parents at the end of 1916, she witnessed her father's distress on finding himself out of work and unwanted, and she regarded his treatment as undeserved and disgraceful. From the mid 1960s, when she first made contacted with Stan Fish and saw hope of restoring McAlpine's standing to its rightful place, until almost the end of her long life (1895–1985), Erica Wedge did all she could to right the wrong done to her father. Her dedication and Stan Fish's persistence played large parts in doing so.

20.3 The Daniel McAlpine Memorial Lectures

The first Plant Pathology conference organised by the Society rather than CSIRO was held in 1974. There was increasing awareness of the significance of Daniel McAlpine's role in the development of plant pathology in Australia, and the appropriateness of the title *Father of Australian Plant Pathology* bestowed on him by Professor L.W. Waterhouse (1939). The Executive of the Society acknowledged this by agreeing to include a *Daniel McAlpine Memorial Lecture,* given by a prominent plant pathologist, at a plenary session of every biennial Society congress thereafter. Dr Lilian R. Fraser (1976) gave the first address entitled, *Diseases of Citrus – the First Hundred Years*[A3]. This was an appropriate title as McAlpine's first treatise on a phytopathological subject was on diseases of citrus found in Australia (McAlpine 1899, 1909). Since then there have been 18 further such dissertations covering a wide range of topics (Appendix 3).

McAlpine would have greatly enjoyed hearing these talks. Some were subjects close to his heart. Dr Dorothy Shaw's (1999) lecture on *Bees and Fungi* would have delighted him as much as it did her audience at the time. He had published several papers on fungi and insects, although none on fungi and bees. John Walker's two lectures[A3] in 1980 and 1991 were perhaps the closest to McAlpine's experience and enjoyment during his 21 years as Victoria's Vegetable Pathologist. Dr Gretna Weste's autobiographical lecture, *A Long and Varied Fungal Foray* would also have appealed since her professional journey had parallels with his own (Weste 2005). He may also have been somewhat amused at Professor Allen Kerr's (1987) comparison of his background with McAlpine's own, each being of Scottish heritage, and each having made their mark in Australia, but he would also have been intrigued by Kerr's lecture entitled, *Agrobacterium: Pathogen, Genetic Engineer and Biological Control Agent*, which for the most part for McAlpine would have opened a window into a brave new world, mostly beyond imagination in McAlpine's day. He would certainly have been delighted that his daughter, Erica Wedge, was a guest of honour at the opening of the Fourth International Plant Pathology Congress held in Melbourne in August 1983. At that congress, Professor Matthew's lecture[A3], *Relationships between Plant Pathology and Molecular Biology*, opened another window on research in plant pathology which, over the past three decades, has had enormous impacts on many facets of our discipline, both in theory and practice, such as understanding processes driving pathogenesis, and in practice, by improving the speed and reliability of diagnosing diseases caused by all biological agents.

In 1990, following a review of agricultural research in Victoria, three awards were instituted, one each for a student, a departmental scientist and a farmer (Edwards 2014). The awards are made to demonstrate the Department's commitment to excellence in the field of agricultural science and to celebrate successes in the continuum from the foundation of scientific principles through to applications in industry. Since the inauguration of the awards in 1992, interest in the nominees and award winners has grown substantially and they are deemed to be serving their purpose in addition to perpetuating the memories of those they honour.

20.4 The Daniel McAlpine Outstanding Achievement Award

The Daniel McAlpine Outstanding Achievement Award is for a departmental scientist or team that has made a significant achievement over the previous 5 years (Edwards 2014). The criteria for selection fall within the fields of research, extension, policy or regulation which enhance the profitability and sustainability of natural resource management or agricultural production, assessed against five criteria: (i) the quality of the science; (ii) the contribution to sustainable economic growth of Victoria's land and resource based industries; (ii) the methods used and any methodological innovation; (iv) the scientific, economic, social and environmental impact, and: (v) effective communication of the science to appropriate stakeholders. The award provides $5,000 for professional development and a uniquely sculptured trophy.

20.5 The Daniel McAlpine Medal

In recent years the International Mycological Association has established awards to encourage promising young mycologists, and to reward extraordinary lifetimes' of service to mycology. The inaugural awards, including cheques for 500 euros were to be made at the 10th International Mycological Conference in Thailand in 2014. The six groups of awards go to young achievers of promise including:

The Keisuke Tubaki Award – Asian Mycological Association;
The Elias Magnus Fries Award – European Mycological Association;
The Carlos Luis Spegazzini Award – Latin American Mycological Association;
The Arthur Henry Reginald Buller Award – North American Mycological Association;
The Ethel Mary Doidge Award, African Mycological Association;
The Daniel McAlpine Award – Australasian Mycological Association.

Thus, Daniel McAlpine has also been saluted with a mycological award in his honour. Established in 2010, *The Daniel McAlpine Award* is sponsored by the International Mycological Association (IMA) and is awarded to an outstanding young mycologist in the early stages of her or his career (within 10 years of receiving their PhD). Only members of the IMA can nominate mycologists for these awards that are bestowed at International Mycological Association Congresses. The citation accompanying the first Australasian award shines a light on the standard aimed for and the appropriateness of the award in respect to the man whose memory and achievements it honours.

The 2010 awarded went to Dr Ceri Pearce for her research and major contributions to the mycological community. PhD research on the biology and taxonomy of Australian fungi, resulted in authorship of a book on the Phyllachoraceae of Australia. She then focused on the increasingly important area of biosecurity and became involved in the development of diagnostic and emergency response systems for exotic pests introduced into Australia. She made major contributions to the mycological community through her roles on mycological councils, which included executive councillor and librarian of the Australasian Mycological Society from 1998 to 2002, executive councillor for the Australasian Plant Pathology Society from 2005 to 2007, and co-chair of the Regional Councillor Working Group of the Australian Plant Pathology Society from 2005 to 2008. In addition she won the bid and successfully co-organised the 8th International Mycological Congress in Cairns in 2006, the first time the congress was held in the Southern Hemisphere. Dr Pearce has also been involved in numerous mycological education and training programs, both within her organization and for agricultural industries, peers and school students.

20.6 Wide Influence and Many Talents

Between 1890 and 1906, when he was helping organise the inter-colonial conferences on wheat rust, McAlpine cooperated closely with William Farrer and did so for many years. Following his withdrawal to *Invercliffe* he continued to correspond with colleagues providing assistance and discussing subjects of mutual interest. Such was the case with W.L. Waterhouse (1922a, b) who within months of his appointment as lecturer in genetics, plant breeding and plant pathology in the School of Agriculture at Sydney University wrote for information on McAlpine's finding on rust resistance in barley and sharing their mutual interest in and observations on the infection of barberry by wheat rust. Scientists in Japan, Scandinavia, the USA and other parts of the world also corresponded with him. While it was Farrer who produced rust-resistant wheat cultivars, such as *Federation* and *Florence*, it was McAlpine who was one of the main drivers of the Rust-in-Wheat Conferences and conducted much of the rust resistance testing. E.C. Large (1940) acknowledged their importance when he wrote; 'The mighty rust investigation began in Australia with a series of Rust-in-Wheat Conferences following the epidemic of 1889. There were Cobb and McAlpine out there, and Farrer, a private wheat breeder who produced some of the earliest improved hybrid varieties. Then in the United States there was Mark Alfred Carleton; in Canada, William Saunders and his sons; and in Sweden, Professor Jacob Eriksson. These were among the men whose names would be remembered'. McAlpine maintained correspondence with most of them.

Those who have enjoyed the scientifically based investigations of Sherlock Holmes and his sidekick Watson might reflect on the likely influence of McAlpine on Holmes' success! The creator of this popular character, Arthur Conan Doyle, corresponded with McAlpine over the years. Doyle had studied medicine at the University of Edinburgh between 1876 and 1881 and would have attended McAlpine's lectures on botany and natural history at the Pharmacy College of the Medical School, where McAlpine was recognised as a popular and influential teacher. It is not too much to suppose that to some extent McAlpine's influence attracted Doyle into the adventures of a forensic scientific investigator, especially as forensic medicine was developing into a science at that time. Doyle maintained a correspondence with McAlpine (Wedge 1984), and on a visit to Melbourne in October 1920, he made contact with him (Doyle 1920) from the Grand Hotel before going on to Sydney for a debate with a society of 'Materialists', Doyle espousing 'Spiritualism' at the time. By then McAlpine and his wife were at Wandin.

Prior to McAlpine's death, W.M. Carne (1927–'28) referred to him as *The Father of Plant Pathology in Australia*, a sentiment publicised by Professor W.L. Waterhouse (1939) when he added that McAlpine's enormous contribution to plant health in Australia and beyond should never be forgotten.

References

Anon. (1932) The late Daniel McAlpine and Nathaniel A. Cobb; the passing of two pioneer plant pathologists. Suaga 4(3):15–16

Carne WH (1927–'28) An outline of the history of phytopathology with special reference to its development in Australia. J Proc R Soc West Aust 13:24–36

Carne WM, Pittman HA, Elliot HG (1929) Studies concerning the so-called bitter pit of apples in Australia – with special reference to the variety "Cleopatra", CSIR bulletin no. 41. Government Printer, Melbourne, 101 pp

Doyle, Sir Arthur Conan (1920) Letter [October 1920] written to McAlpine from the Grand Hotel, Melbourne

Edwards J (2014) An email [5-5-'14] containing information in the Department of Primary Industries Science Awards sent to the author

Fish S (1949) Letter written [7-6-'49] to Dr. C.J. McGee regarding commemoration of McAlpine's birth year

Fish S (1966) Letter [8-6-'66] written to F.C. Kent, Editor, Australian Journal of Pharmacy

Fish S (1970) The history of plant pathology in Australia. Annu Rev Phytopathol 8:13–36

Fish S (1976) Daniel McAlpine – a pioneer plant pathologist in Australia. Aust Plant Pathol Newslett 5(1):11–14

Fraser L (1976) Diseases of Citrus trees in Australia – the first hundred years. In: Daniel McAlpine memorial lecture, second national plant pathology conference, May 1976, Brisbane, QLD, Australia

Kerr AR (1987) *Agrobacterium*: pathogen, genetic engineer and biological control agent. Aust Plant Pathol 16:308–316

Large EC (1940) The advance of the fungi. Jonathan Cape, London, 488 pp

McAlpine D (1899) Fungus diseases of citrus trees in Australia and their treatment. Department of Agriculture Victoria, Govt Printer, Melbourne

McAlpine D (1909) A disease of citrus caused by *Phoma citricarpa* a fungus which is unknown elsewhere. In: Agricultural Gazette of NSW, Government Printer, Sydney, p 228

McAlpine D (1910) Twenty years of plant pathology in Australia. Report of the vegetable pathologist for 1907–1910. Department of Agriculture, Victoria, pp 56–61

McGee CM (1949) Reply sent [17-6-'49] to S. Fish proposing a commemorative Stamp to celebrate the birth of Daniel McAlpine in 1849

Purss GS (1994) History of the Australasian plant pathology society. Aust Plant Pathol 23:122–128

Shaw DE (1999) Bees and fungi. Aust Plant Pathol 28:269–282

Sissons ATS (1968) History of the Victorian College of Pharmacy. Aust J Pharm 49:497–500

Waterhouse LW (1922a and b) Letters [8-4-'22 and 28-4-'22] written to McAlpine regarding rust resistance in barley lines and interest in rust infection of barberry

Waterhouse LW (1939) Proceedings of the 24th ANZAAS (Australian and New Zealand Association for the Advancement of Science) conference held in Canberra, January 1939, vol 24, pp 234–259
Wedge E (1984) Letters [4th to 7th May-1984] to D.G. Parbery, La Trobe Collection
Weste GM (2005) A long and varied fungal foray. Aust Plant Pathol 34:433–437
Willis JH (1949) Daniel McAlpine; in botanical pioneers in Victoria. Vic Nat 66:83–89

Appendices

Appendix 1

Biographical Notes on Some Who Touched on or Who Influenced McAlpine's Career

Since many of those included in Appendix 1 may not be known to readers, their backgrounds and their relevance to McAlpine's story have been provided in these brief biographical notes. The notes also avoid the tedium of repeating the positions held by those frequently referred to throughout sections of the text, or who impacted on some aspect of the McAlpine story. In several instances, the notes provide an insight into their relationships with McAlpine. There are some players of historic note, who in some way or another, found their way into the story, but who may not have been known to McAlpine. I am also mindful that some readers, especially youthful readers and those from northern climes, may not be familiar with many of the influential and notable Australian personalities of McAlpine's times, but who played a role. Where no reference is given for the entry, either they are readily found on the 'Web', or the entry was compiled from my own digging.

The notes in their totality also draw attention to certain points of social history pertinent to the period. Over half those who built careers in Australia, including McAlpine, were born in England or Scotland. At least three who contributed well to science, Benson, Brittlebank and French Snr, had no formal tertiary education, while at least three others, Baron von Müller, Bancroft and Farrer, migrated to a warmer drier climate for the sake of their health, usually in the hope of limiting the ravages of tuberculosis.

In some cases, for example those of Professor W. Farmer who was involved in the Rust-in-Wheat conference of 1890, and Mr R. Scott, Victorian Government Chemist in 1911, I have been unable to discover their backgrounds.

D.G. Parbery, *Daniel McAlpine and The Bitter Pit*,
DOI 10.1007/978-3-319-09552-3

NB: ADB = *Australian Dictionary of Biography.*
Superscript 'A1' following a name in the text directs the reader to the entry in Appendix 1.

Adam, Dr D.B. – (1900–1952) – Adam graduated B.Agr.Sc from the University of Melbourne, having studied plant pathology under Ethel McLennan. By 1933 he was Head of Plant Pathology in the Victorian Department of Agriculture. He moved to head the Department of Plant Pathology at the Waite Research Institute of the University of Adelaide in 1934 (V.A. Edgeloe, The Waite Agricultural Research Institute: the first 50 years 1924–1974, 1984; *Australasian Plant Pathology* **18**: 47–56. 1989).

Bailey, Frederick Manson (1827–1915) – Bailey was born into a horticultural family in Hackney, London, where he went to school until he was 12 years old. In 1839, the family migrated to Australia, where his father rented a government farm before being appointed Government Botanist in 1841. Bailey had no further formal education, but being an avid reader educated himself. After a brief period running a plant nursery with his father and brother he tried his luck on the Bendigo goldfields. Then in 1858, following his marriage, he took up farming in Lower Hutt, New Zealand. In 1861 he returned to Australia and opened a native plant seed store in Brisbane, supplying overseas markets. In 1875, a Government Board appointed to look into cases of plant poisoning of livestock and plant diseases engaged Bailey as botanist. He travelled widely in his quest for poisonous pasture plants, collecting as he went. Between 1874 and 1881 he published two books on ferns and one on Queensland grasses. In 1881 he was appointed government botanist, and retained the position until his death. During his tenure he published extensively, including a six-volume Flora of Queensland. [*Australian Dictionary of Biography* [*ADB*] **3**: 73–74.]

Bancroft, Joseph (1836–1894) – Bancroft was born and grew up in Manchester England where he studied at the Manchester Royal School of Medicine and Surgery, taking out several prizes. He graduated in 1859, then gained an M.D. degree from St. Andrew's University in Scotland before working at St James Hospital, Nottingham. About 1863 he was advised to move to a warmer climate for his health's sake and in 1864, he arrived in Brisbane on the 523-ton paddle steamer *Lady Young*. There he went into medical practice, which he maintained until his death. He was a keen naturalist and undertook several research projects, which yielded valuable benefits. He was esteemed as a valuable citizen. [*ADB* **3**:84–85].

Benson, Albert H. (1864–1930) – Albert Benson was born in Taunton, Somerset, England; educated at Taunton College and the Royal Agricultural College Cirencester. From 1883 he managed a large estate in Scotland. In California he gained experience in orchard management, and at an American experiment station. In 1894 he migrated to New South Wales and was appointed fruit expert to the NSW Department of Agriculture. His expertise was used to establish experimental farms at Wagga Wagga, Pira Bori and Wollongbar, as instructor at the latter. In 1896 he was appointed Instructor in Fruit Culture for Queensland until 1910 when he was appointed Director of Agriculture for Tasmania. He represented Tasmania on the Bitter Pit Advisory Committee, returning to Queensland at the end of 1915 as Instructor in Fruit Culture for the State, where his expertise in the control of diseases in fruit and fruit trees was greatly valued. He retired at the end of March 1927. [*Brisbane Courier*, 18-10-1930, p. 15]

Berkeley, Rev. Miles Joseph (1803–1889) – Berkeley was an Englishman, Educated at Christ's College Cambridge University. After taking Holy Orders he began his career as an Anglican priest in 1837. By that time he was fascinated by cryptogrammic botany, and, among other groups, he described some 6,000 species of fungi. He made an extensive study of the biology of *Phytophthora infestans*, including the infection of its host and he examined many other important plant pathogens. He is regarded as the pioneer of plant pathology in the United Kingdom [1911 *Encyclopaedia Britannica*].

Brittlebank, Charles Clifton (1862–1945) – Brittlebank was born in Derby, England in 1862, and died in Melbourne on 2nd November 1945. He arrived in Victoria in 1893, where he and his brother Thomas established a dairy farm at Myrniong. His interests included geology, ornithology and botany. He was a member of the Field Naturalist's Club of Victoria and a talented illustrator of biological subjects. He illustrated C. French's *Destructive Insects of Victoria* (1891–1911, 2013) and A.J. Campbell's *Nests and Eggs of Australian Birds* (1900) while also providing illustrations for several books and bulletins by Daniel McAlpine. Brittlebank discovered ice-age evidence in the Werribee Gorge ca 1894. He was assistant to D. McAlpine from 1908, and succeeded him to become government plant pathologist of Victoria 1913–24; he was biologist in charge of the Science Branch, Victorian Department of Agriculture 1924–28; lectured on fungal diseases of fruit at the Burnley School of Horticulture 1923–28; lectured on plant pathology, at the School of Agriculture, University of Melbourne ca 1920–1928; prepared an index of Australian fungi and fungal diseases for the Council for Scientific and Industrial Research. [*Encyclopedia of Australian Science*].

Brown, Professor William (1888–1975) – One of the first to become involved in the study of the physiology of parasitism and, from 1912, Professor of Plant Pathology at the Imperial College of Science and Technology, London.

Butler, Sir Edwin John FRS (1874–1943) – Butler was born and educated in Ireland, where in 1898 he graduated from Queen's University, Cork with the degrees M.B, B.Ch and B.A.O. Later he received a M.Sc. in Botany from University College Cork in 1920, as well as other degrees and honours. In the meantime he studied mycology in Paris, Antibes, Freiburg and Kew. In 1900, on the recommendation of the Director of the Royal Botanic Gardens, Kew, he was appointed to the Government of India as the Cryptogamic Botanist. Over the following two decades he gained wide experience as a mycologist and plant pathologist, and as Director and Principal of an Agricultural College in Pusa. In 1920 he returned to England as foundation Director of the new Imperial Bureau of Mycology (IBM, later CMI), Kew, which was established to provide advice on plant diseases throughout the British Empire.

He was widely respected, and perhaps best known to my generation of plant pathologists through his excellent textbook *Plant Pathology,* written with Dr S.G. Jones and first published by Macmillan Press in 1949. [*Oxford Dictionary of National Biography*].

Cameron, Dr Samuel Sherwyn (1866–1933) – Cameron was Chief Veterinary Inspector with the Victorian Department of Agriculture from 1905. His notes and signature on Department of Agriculture correspondence prior to 1911 suggest he was appointed Acting Director of Agriculture from about 1909. In 1910 the Victorian Government reorganised the Department of Agriculture and in January 1911 appointed him Director of Agriculture. He retired in 1931. He supported the establishment of the Faculty of Agriculture at Melbourne University in 1906 and lectured there for many years. He provided strong support to McAlpine during the bitter pit investigation. [*ADB* **7**: 539]

Carmody, P. J. – Carmody was Chief Inspector of Orchards for the Victorian Department of Agriculture who represented the interests of Western Australia on the *Commonwealth Advisory Committee on Bitter Pit.*

Cherry, Professor Thomas (1861–1945) – Cherry began as a medical student at Ormond College in 1885. He was appointed a lecturer in Microbiology at the University of Melbourne in 1902. In 1905 he was appointed Director of Agriculture for Victoria until replaced by Dr Cameron, after which he was appointed to the Chair of Agriculture in 1911 until funding for the Chair was withdrawn. Between 1911 and 1916 he served a term as Dean of the Faculty of Agriculture at Melbourne University. [*ADB* **7**: 633–634].

Cobb, Nathan Augustus (1859–1932) – Cobb was born in Massachusetts, USA, obtained a PhD from the University of Jena, Germany before working at the Naples Zoological Station. In 1889 he migrated to Sydney with his family and in 1890 took up an appointment as temporary Professor of Biology at the University of Sydney. Later that year he was appointed Vegetable Pathologist in the NSW Department of Agriculture. He resigned from the post on 5th May 1905 and returned to the USA. [*ADB* **8**: 39–40].

Cook, Hon. Sir Joseph (1860–1947) – Cook was born in Silverdale, Staffordshire England where in 1873 his father was killed in a mining accident. After supporting his mother and siblings for a decade he migrated to Lithgow in New South Wales where he worked at the coal mine, became an executive member of the miners' league through which 2 years later he entered politics as a member of the Labour Party. By the early post federation period he was increasingly supportive of conservative ideals and by 1905 was deputy-leader of the Free Trade Party in the Federal government. He remained in Parliament during the turbulent period of the following decade often in opposition and then in May 1913. Five months after the resignation of Alfred Deacon he was elected Prime Minister and served until October 1916 when Billy Hughes was elected Prime Minister, with Cook as his deputy. Cook was a colourful, strong minded character. [*ADB* **8**: 96–99].

Cooke, Mordacai Cubitt (1825–1914) – Cooke was born in Norfolk, England where he worked initially as a fabrics merchant then a clerk in a law firm, but his main interest was botany. He taught Natural History in Lambeth, after which he became curator of the India Museum at India House in London from 1860 to 1979. When its botanical collection was transferred to Kew, he went with it. There he became absorbed in mycology. Being especially interested in Australian fungi, he was commissioned to write a *Handbook of Australian Fungi* by the Colonial Governments. He lent McAlpine many fungal collections made by early Australian explorers and collectors. [See English, M.P. (1987) Chapter 6].

Cumpston, Dr John Howard Lidgett (1880–1954) – A distinguished university graduate in Medicine in 1902, he became committed to preventive and public medicine. From 1908 his main interest was in disease abatement and he joined the Federal quarantine service as general quarantine officer in Western Australia in November 1910. In August 1911 he was appointed chief quarantine officer in Victoria. In September the director, W. P. Norris, travelled overseas, leaving Cumpston in charge of the Federal service until May 1912. At the end of that year Cumpston moved to Brisbane as quarantine supervisor in Queensland. Norris transferred to London, and Cumpston returned to Melbourne in May 1913 as acting director, being confirmed in that post from 1st July 1913. He replaced Norris as chairman of the Advisory Committee on Bitter Pit. [*ADB* **8**: 174–176].

Darnell-Smith, George Percy (1868–1942) Darnell-Smith was born in Chipping Norton, England. After graduating B.Sc. (Hons) from the University of London he lectured at the Merchant Venturers' Technical College Bristol from 1893 to 1907 and was made a fellow of the (Royal) Institute of Chemistry of Great Britain and Ireland. From 1907 to 1909 he was Assistant Director of Technical Training in New Zealand and then in April 1909 he was appointed Assistant Microbiologist at the new Bureau of Microbiology in Sydney on a salary of £500 per annum. In 1913 the Bureau was incorporated into the NSW Department of Agriculture where he inaugurated the Biology Branch, retiring as biologist in 1927 and as Director of Agriculture in 1930. Among his many valuable contributions to Agriculture was his development of copper carbonate powder, used worldwide as a wheat seed dressing. He was a member of the Commonwealth Advisory Committee on Bitter Pit. [*ADB* **8**: 212–213]

Deacon, The Hon. Alfred (1856–1919) – Deacon was born in Melbourne where he took a Law degree from Melbourne University in 1877. From 1879 to 1890 he was a member of the Colonial Government of Victoria. He was an able administrator and in 1884 held the portfolio of water supply and irrigation. Following a visit to California he persuaded the Chaffey Brothers to leave California and establish an irrigation scheme in Mildura in 1886. Deacon was responsible for the appointment of McAlpine as Vegetable Pathologist. Between 1903 and 1910 he served several terms as Prime Minister of the Federal Government of Australia. [*ADB* **8**: 248–256].

De Bary, Heinrich Anton (1831–1888) – Anton de Bary was born and educated in Frankfurt. In 1850 he qualified as a medical practitioner in Berlin after which he practiced briefly in Frankfurt before joining the University of Tübingen. Two years later he moved to Freiberg and in 1867 accepted the Chair of Botany at Halle University. In 1871, following the Franco-Prussian War, he moved to the University of Strasbourg. De Bary is regarded as the father of mycology (Ainsworth 1984) but he also provided strong plant pathological evidence (de Bary 1853) in support of Rev. M.J. Berkeley's contention that a microscopic fungus was the cause of late blight of potato. Between 1866 and 1884 he prepared a text on the comparative morphology and biology of fungi etc. (de Bary 1887), which launched mycology into the modern era and secured de Bary's place as the father of modern mycology (Ainsworth 1976) [See References Chapter 5].

Donalson, Professor Sir James, LLD (1831–1915) – Donaldson was a classical scholar, educated at the University of Aberdeen, New College London and in Berlin. In 1852 he was appointed assistant to John Stuart Blackie, Professor of Greek at the University of Edinburgh, then in 1854 he was Rector of Sterling High School. In 1866 he became Principal of the Royal High School Edinburgh, then in 1881 Professor of Humanities at Aberdeen University from 1881 to 1886. He was then appointed Principal of the United College of St Salvador and St Leonard in the University of St Andrews, where in 1889 he was appointed Vice-Chancellor of the University, discharging his duties until a few days before his death. He was Chairman of the Board of Directors of the Watt Institute and School of Arts, during Daniel McAlpine appointment from 1875 to 1884 [*Encyclopaedia Britannica*, 11th Edition].

Etheridge, Professor Robert (1819–1903) – Etheridge was a student of natural history, well versed in botany, geology and palaeozoology. From 1865 he was assistant to Professor Huxley in the preparation of *Catalogue of Fossils in the Museum of Practical Geology*. He was highly regarded by his peers and in 1880 was President of the Geological Society of London. His son, Robert Etheridge (1846–1920) was appointed Assistant Geologist for the Geological Survey of Victoria, Australia, from 1866 to 1869. Robert Jnr returned to the UK in 1871, then in 1887 returned permanently to Australia where he continued geological surveys and in 1895 was appointed Curator of the Brisbane Museum, pursuing his interest in and publishing notes and papers on Australian and New Guinean geology and ethnology [*ADB* **8**: 442–443].

Ewart, Alfred James (1872–1937) – Ewart was born at Toxteth Park, Liverpool, England on 12th February 1872. His father, a Scot, was lecturer in chemistry and director of the chemistry laboratory at the Liverpool Institute and School of Art. Ewart studied science at University College Liverpool from 1890 to 1893, graduated from the University of London with a B.Sc. (Hons), and was appointed demonstrator in botany. His first class honours degree was granted for research into seed germination for which he was awarded the 1851 Exhibition Scholarship. This enabled him to study under Wilhelm Pfeffer at the University of Leipzig where he gained his PhD in 1896, and where he studied the inhibition of photosynthesis through the accumulation of synthate. His success in Leipzig resulted in his scholarship being extended, allowing him to travel and study in Indonesia, Sri Lanka and Singapore. From the time of his graduation until his return to England in 1897 he published a number of notable papers, which gained him the award of a D.Sc by the University of London. During the following 9 years he occupied a number of academic positions in England. From mid-1898 he was an extension lecturer and research scholar at Oxford University where he translated Professor Pfeffer's classic three volume *Physiology of Plants* from German into English. His published work continued to gain recognition and in 1910 he was awarded a D.Sc, from the University of Oxford. In 1922 he was made a Fellow of the Royal Society of London.

During preparation of the University Act of 1904, the Victorian Premier, Thomas Bent, conscious of the importance of agriculture to the state economy, urged the establishment of courses in agricultural science at the University of Melbourne. One consequence was that the University Council established Australia's first Chair of Botany specifically to teach agricultural science students. Baldwin Spencer, Professor of Biology, suggested the new professorship be joined with the vacant position of Government Botanist. As was then the custom, the selection of a suitable appointment was made in London. The selection committee comprised Professor H. Marshall Ward (University of Cambridge) and Professor Francis Oliver (The University of London). They selected Alfred Ewart.

Each of Ewart's appointments was predominantly for support of the State's agriculture. At the university his primary responsibility was to teach botany to agricultural science students, while at the Herbarium, as Government Botanist, his primary responsibility was servicing the needs of agriculture. His role included the identification of, and the supply of appropriate information about, plants of agricultural importance, including weed species and noxious plants, their biology and control. He was also required to assist with identifications and enquiries and provide advice for Customs and Quarantine. He took his responsibilities seriously and expanded his brief as appropriate. One example was his restoration of order to the Herbarium, some aspects of which had been largely neglected over the decade since von Müller's death. His earlier research into seed biology made him aware of the need for weed-free batches of viable seed for agricultural and other purposes. As a result of his vision and efforts, Commonwealth and State seed-testing laboratories were established. Because he found that the reliability of plant identifications and advice provided by some government departments and branches was often erroneous or inappropriate, he insisted that identifications and advice relating to plants in any government publication had to be signed off by him. Any public servant who failed to follow his edict was left in no doubt about what was required in future. In 1921 his appointment at the University was made fulltime [*ADB* **8**: 448–450; *Historical Records of Australian Science* **16**: 138–167].

Ewart, James Cossor FRS (1851–1933) was born in Pennicuik, Midlothian, Scotland, educated at the University of Edinburgh, and was a highly regarded zoologist. He was Professor of Natural History at Edinburgh from 1882 until 1927. No connection with Professor A.J. Ewart has been established, although Alfred Ewart's father was also born in Scotland, a few years after J.C. Ewart. [*Nature* **133**: 165–166, 1934].

Farmer, Sir John Bretland MA, D.Sc., FRS (1865–1944) was Professor of Botany at the Imperial College of Science and Technology, South Kensington. He was interested in plant chemistry and worked on bryophytes at the Chelsea Physic garden [*Obituary Notices of Fellows of the Royal Society* **5** (14): 17, 1945].

Farrer, William James (1845–1906) – William Farrer was an Englishman born into a farming family in Docker, Lancashire. He was bright and won scholarships which took him to Christ's Hospital, London, then to Pembroke College, Cambridge where he gained a B.A. in 1868. He began studying medicine, but shortly afterwards contracted tuberculosis so in 1870 migrated to Australia for the sake of his health. His first job was tutor on a sheep station, *Duntroon*, now part of Canberra, where he took an interest in agriculture. After qualifying as a surveyor he joined the Lands Department, which took him over much of NSW including large areas of the developing wheat belt. During this period, 1875 to 1876, he concluded that the fundamental problems with the wheat industry was that English wheats were not suited to the harsh Australian environment, so he put his mind to devising a selection and cross breeding plan in order to improve matters. Gregor Mendel's discoveries of the genetic basis of inheritance, published in 1866 but not brought to public notice until 1900, were not available to Farrer, who relied on information obtained from European and American scientists interested in plant improvement through breeding. He resigned from the Lands Department in 1886, settled on a property, *Lambrigg*, where Canberra now stands and applied himself to his objective. His initial success came through crossing selected local wheats with drought-resistant Indian wheats. He then became aware of the much better milling and bread making flour quality of Canadian wheats. In order to select the best parental lines for flour quality and their progeny he needed the help of a chemist which led him into a long term partnership with Frederick Guthrie, who in 1892 devised procedures for simulating milling and baking procedures. In 1898 Farrer was employed by the NSW Department of Agriculture as Plant Breeder at a salary of £350 per annum. In 1890, Farrer met Daniel McAlpine, whose assistance with the pathology aspects of breeding for resistance to rusts, bunts and smuts was also of critical importance to Farrer's great success [*ADB* **8**: 471–473].

Fisher, Hon Andrew (1862–1928) – Fisher was a Scot from the coalfields near Kilmarnock, who migrated to Queensland where he was active in labour politics and entered the Queensland Colonial Government in 1893. During various periods between 1908 and 1915 he served as Prime Minister of Australia. [ADB **8**:502–507].

Fisher, Sir Ronald FRS (1890–1962) – Fisher joined the staff at Rothamsted in 1919, where he began a major study of the vast amounts of data accumulated from field experiments at Rothamsted, some running since the mid-1840s. In 1925 he published *Statistical Methods for Research Workers*, followed in 1935 by *The Design of Experiments*. These two books revolutionised experimental design and the extent to which research outcomes could be relied on with respect to numerical differences. In later life he moved to Adelaide where he continued his work for some years [*Encyclopaedia Britannica*].

Flentje, Prof. Noel Thomas (1921–1974) – Flentje was born in Rochester, Victoria. He graduated B.Agr.Sc. from Melbourne University in 1943 where he studied plant pathology taught by Dr Ethel McLennan. He joined the staff of the Waite Agricultural Research Institute, University of Adelaide, in 1944 and in 1964 was appointed head of the Department of Plant Pathology [*Encyclopaedia of Australian Science; Journal of the Australian Institute of Agricultural Science* **40** (3): 206, 1974].

French, Charles (1842–1933) – Charles French was born in England where he developed an interest in collecting and naming insects. In 1852 the family migrated to Victoria where he eventually found his way into horticulture under von Müller and then Guilfoyle who transferred him from garden duties to the herbarium where again he was working for von Müller and where by 1886 he had risen to first herbarium assistant. In 1889 he was made Government Entomologist. Between 1891 and 1911 he published Volumes I–V of *Handbook of Destructive Insects of Victoria*. The sixth volume was prepared but for some reason not published. In time it was believed lost. In 2011 it was found with the McAlpine files and was published in 2013 [*ADB* **8**:585–586].

French, Charles Hamilton (1868–1950) – Charles French Jnr followed his father's career path in many respects and published many articles on insects. In 1896 he was appointed to the Victorian Department of Agriculture under the Vegetable Diseases Act. From 1931 until his retirement in 1933 he was Biologist in the Department of Agriculture and Officer-in-Charge of the Science Branch at Burnley Gardens [*ADB* **8**:585–586].

Geikie, Sir Archibald (1835–1824) – A Scottish geologist and graduate of the University of Edinburgh, Geikie made a geological survey of Scotland and assisted in the geological survey of Great Britain. He was appointed to the first Chair of Geology and Mineralogy at the University of Edinburgh in 1871. In 1881, he succeeded Sir Andrew Ramsay as Director of the Museum of Practical Geology in London, and at times, lectured at the Royal School of Mines. His younger brother James Geikie (1839–1915), also a highly respected geologist, succeeded Sir Archibald as Professor of Geology at Edinburgh University in 1882. McAlpine corresponded with Sir Archibald throughout his life and also occasionally with James [*Encyclopaedia Britannica*].

Gilbert, Sir John Henry (ca 1816–1901) – Gilbert, a chemist, and Sir John Bennet Lawes, owner of the sixteenth century estate *Rothamsted Manor* and producer of the first artificial fertiliser for agricultural use in 1842, founded the Rothamsted Experimental Station in 1843. Over the next 57 years they established the foundations of modern British agriculture and the principles of crop nutrition [*Encyclopaedia Britannica*].

Graham, George – (1838–1822) – A progressive farmer in the Shepparton District, Graham was Minister for Water Supply and for Agriculture intermittently in various Victorian Governments between 1890 and 1913. [*ADB* **9**:67].

Grant, Sir Alexander (1826–1884) – Grant was a British educationalist who in 1868 was appointed Principal of the University of Edinburgh. He was largely responsible for the huge success of the Tercentenary Celebrations of the University. McAlpine gave him a copy of the first volume of his Botanical Atlas, which Grant presented to the library. It is possible that he made a verbal offer of a D.Sc. to McAlpine, if indeed anyone did [1911 edition, *Encyclopaedia Britannica*].

Grasby, William Catton FLS (1859–1930) – Grasby was born in South Australia and from an early age was deeply interested in education. In 1881 he made a study tour of Europe, returning to teach in state primary schools where he introduced libraries and nature museums. He experimented with nature study, drawing and science and founded the Boys' Field Club for Saturday nature excursions. He became highly critical of Education Department policies. In 1889 he toured North America, Britain and Europe, studying educational innovation and published his findings 1891. He stirred debate on teacher training, free education in secondary schools and agricultural colleges, and changes in school building and furniture design. He edited the 12 issues of the journal *Educator* (1893–94). From 1892 to 1894 he was director of agronomy and manual training at Way College, Unley. From 1884 to 1886 he taught at Roseworthy College. In 1896 he toured Mediterranean countries as an honorary Government Commissioner reporting back on fruit growing techniques, after which he returned to Way College for 2 years. He spoke and wrote widely on educational and agricultural matters. From 1896 to 1904 he owned and edited *Australian Garden and Field*, from 1905 he was the agricultural editor of the *Western Mail* and *West Australian*, gaining a huge following for his weekly advice column for farmers, entitled *Mutual Help*. He continued to teach agriculture in high schools and in 1912 published an important textbook, *Principles of Australian Agriculture*. He advocated tertiary education for Western Australia and from 1912 to 1914 was a member of its University's first Senate. He was a fellow of the Linnean Society of London, and president of the Royal Society of Western Australia during 1929–30. [*ADB* **9**:81–82].

Guilfoyle, William Robert (1840–1912) – Guilfoyle's father was an English landscape gardener and nurseryman. The family lived in Chelsea when William was born, but migrated to Sydney in 1853 where William was educated and developed his interests in plants and landscape gardening. He was a member of the scientific staff aboard HMS *Challenger* on its voyage of exploration around the Pacific Ocean in 1868, after which he settled in northern NSW to farm tobacco and sugar cane, and where he met Ferdinand von Müller. This meeting led to his appointment as director of the Botanic Garden in Melbourne in 1873. Whereas the previous directors had developed the gardens on scientific lines, Guilfoyle placed more emphasis on an aesthetic approach, which defines the gardens to this day. [*ADB* **4**:307–308].

Guthrie, Frederick Bickell (1861–1927) – Guthrie was born in Mauritius and educated as a chemist in England and Germany. He demonstrated in chemistry at Queens College, Cork, Ireland, from 1882, and then from 1888 at the Royal School of Science, South Kensington in London. He was appointed to a similar position at Sydney University in 1890, where he also served as Acting Professor for three periods between 1896 and 1909. In 1892, however, he was appointed chemist to the newly founded Department of Agriculture in New South Wales, retiring in 1924. His first job was to help William Farrer assess the baking quality of the wheats he was using as parental lines and their progeny [ADB **9**: 143–146].

Haggard, Sir Henry Ryder KGB (1856–1925) – After his secondary education, Haggard spent 2 years in London where he frustrated his father's attempts to encourage him into an army or colonial service career. Instead he studied psychic phenomena. Following 7 years volunteer service with the British Colonial Administration of South Africa he returned to England, studied law and was called to the bar in 1884. Not much interested in the law, he turned to writing novels such as *She*, *Ayesha*, *King Solomon's Mines* and many others, which became very popular and for which he was possibly best known. He also became deeply interested in agricultural reform throughout the British Empire. He studied agriculture widely in Europe and the colonies as a layman, and wrote factual books and newspaper articles on world agriculture [*Encyclopaedia Britannica*].

Hagelthorn, The Hon. Frederick William (1864–1943) – Hagelthorn was Minister for Agriculture under the Watt and Peacock administrations from November 1915 to November 1917. He oversaw a great expansion of fruit growing in the state and concerned himself with the provision of infrastructure necessary to support it. He worked closely with Dr Cameron, Professor Cherry and Dr A.E.V. Richardson in regard to water supply. He championed legislation benefiting agriculture, including that relating to closer settlement, railways and water supply [*ADB* **9**:155–156].

Hughes, The Hon. William Morris (1862–1952) – When The Hon. Andrew Fisher resigned as Prime Minister in October 1915, Hughes succeeded him. Prior to 1915, he had served as Acting Prime Minister as required from time to time from prior to the establishment of the investigation in to bitter pit of apples. A man of small stature but great charisma, he was affectionately known as 'Billy' Hughes. He was a member of the British Imperial War Cabinet from during 1916 to 1918 during which period he tried twice to win electoral approval to bring in conscription so that he could provide more troop so that those in France could be relieved more readily [*ADB* **9**:393–400].

Hutchinson, The Hon. William (1864–1924) – Hutchinson was born at Pleasant Creek near Stawell in Victoria. His father was a miner and his mother an aunt of H.V. McKay, founder of the *Sunshine Harvester Company*. It was this cousin who developed the combine harvester used by McAlpine for harvesting experimental wheat plots. Hutchinson entered Parliament in 1902 and was Minister for Water Supply and Agriculture from 1913 to 1915 when he became Minister for Lands [**ADB 9**:414–415].

Huxley, Thomas Henry FRS (1825–1895) – Thomas Huxley was born in London. He received little formal education or systematic training in science, not rare at that time among those who made their names in science. At the age of 16 he undertook a medical apprenticeship which won him a scholarship to Charing Cross Hospital. In 1846 he joined the naval medical service and was assigned to *HMS Rattlesnake* as assistant surgeon and naturalist, sailing on a survey of Australia's north eastern waters. During this time he confined his scientific interest to the comparative anatomy of mollusca and other marine invertebrates. His studies of comparative anatomy transformed understanding of the morphology of the animal kingdom and gained him election as a Fellow of the Royal Society in 1849. He met his future wife in Sydney, explored parts of eastern Australia from Cape York to Hobart and visited New Guinea in 1849 where he developed an interest in anthropology. In 1850 *Rattlesnak*e returned to the UK, where Huxley applied for Chairs at several

Universities, but without success. During his time in Sydney, the topic of establishing a University was widely discussed and Huxley set his heart on appointment to the Chair of Natural History when the time came. Sydney University was founded in 1850. In 1854, Huxley retired from the navy and joined the Royal School of Mines as lecturer and naturalist and was naturalist for the geological survey of Britain in 1855, but he greatly valued his association with Australia and kept reminding friends in Sydney not to forget him and his interest in an appointment. He strongly supported Charles Darwin's theory of evolution, becoming known as 'Darwin's bulldog'. [*ADB* **1**:577–578].

Kühn, Julius (1825–1910) – After leaving school in 1839, Julius Kühn worked on his father's and other farms until 1855. He then began formal training in Bonn, and took out a doctorate from Leipzig. He then took up estate management for a period before being appointed Professor of Agriculture at the University of Halle in 1862. Early on, Kühn developed an interest in diseases of beet and rape, which became the subject of his doctoral thesis. He produced 70 papers on plant diseases as well as papers on other aspects of agriculture. His book, (Kuhn, 1858) *Diseases of Cultivated Crops, Their Causes and Prevention*, based mostly on his own experience and research, proved very popular. In 1859 it was republished in two parts, the first discussing general aspects of plant diseases and the second devoted to specific diseases of common crops [G.C. Ainsworth, *Introduction to the History of Plant Pathology*, 1981].

Laidlaw, William (ca 1866–1925) – Laidlaw, a Scot graduated B.Sc. from the University of Edinburgh after which he worked in Canada before migrating to Victoria in 1908. He soon obtained an appointment in the Department of Agriculture where he worked in horticulture and later, in plant pathology. In 1911 he was appointed Government Biologist but following a reorganisation of the Department in 1914 he was appointed Chief Biologist. In January 1924 he was appointed Director of the Melbourne Botanic Gardens and Government Botanist. He died suddenly at the Gardens in September 1925 [R. Pescott, *The Royal Botanic Gardens*, Melbourne 1981].

Lodge, Sir Oliver Joseph (1851–1940) – Lodge was born in Staffordshire, England, attended a local dame-school and had a somewhat Dickensian education at Newport Grammar School in Shropshire. He developed an early interest in science. In 1872 he gained entry to the Royal College of Science under Huxley, Frankland and Guthrie, where Archibald and Daniel McAlpine were his contemporaries. After hearing J.C. Maxwell speak at a British Association meeting, Lodge bought Maxwell's *Treatise on Electricity and Magnetism.* His interest in the nature and properties of electricity developed from there. By 1894 he was investigating radio waves and telegraphy and 2 years later turned his attention to X-rays and atoms, making a pioneering application when he applied X-ray to locate a bullet in a boy's hand. In 1900 he was appointed principal of the new University of Birmingham where he took up the exploration of radio. He was a popular broadcaster and speaker, greatly honoured and admired [*Oxford Dictionary of National Biography*, Oxford University Press 2004].

Lowrie, William (1857–1933) – Lowrie rose from a humble Scottish agricultural background until in 1883 he obtained an M.A. degree from the University of

Edinburgh. In 1884 he won a Highland and Agricultural Society Scotland bursary, which enabled him to return to the University of Edinburgh to studied agriculture and related sciences. He graduated B.Sc in 1886, with excellent first class result. After appointment at Gordon College, Aberdeen, he was appointed Professor of Agriculture and principle of Roseworthy College in South Australia. In 1901 he departed for New Zealand where he was appointed Director of Canterbury Agricultural College. In 1908, he accepted the position of Director of Agriculture for Western Australia. Sir Newton Moore was his brother-in-law [*ABD* **10**: 160–161].

McAlpine, Archibald Nichol (c.1850–1924) – Professor Archibald McAlpine was Daniel's youngest brother. He was born in Saltcoats but as a young boy was sent to Dumbarton where he attended the Free Church School, and where he later pursued his studies at the Dumbarton Academy where he also became a junior member of staff. He won a scholarship which took him to the Royal School of Mines in London. Two of his class were students who became Lord Herbert Horatio Kitchener and Sir Oliver Lodge. He took out a B.Sc with honours in botany and chemistry and won an exhibition which enabled him to study mathematics in Dublin. However, there he was drawn into a love of botany by Professor M'Nab and admitted as an Associate of the Royal College of Science for Ireland. Then, in Germany he studied botany under Sachs and Sempel in Leipzig and Wertsberg. He returned to Scotland in 1884 to take over the courses Daniel had taught at the Watt Institute and the Veterinary College before joining the staff as lecturer in agricultural botany at the Technical College in Glasgow, on which the West of Scotland College of Agriculture was established under Sir Robert Patrick Wright and remained there until his death. He was a brilliant lecturer, loved by his students for his wit and humour and the fanciful way he presented science [Obituary; *The Lennox Herald* 13th December 1924].

McCoy, Sir Frederick (1817–1899) – McCoy was born in Dublin where he began a study of medicine, then Natural History where his interest focused on Geology and paleontology through which he built up a high reputation. In consequence he was one of the four foundation professors appointed to the University of Melbourne in 1854. He was an anti-Darwinist, energetic and colourful, and contributed fully to the scientific life of the colony. By 1884, however, students were highly dissatisfied with his lectures and colleagues were becoming increasingly critical of him. McAlpine presented him with copies of his Botanical Atlases, which were later acquired by Dr. Sophie Ducker and presented to the Melbourne Royal Botanical Gardens Library. McCoy retired in 1887. [*ADB* **5**:134–136].

McFarland, Sir John Henry (1851–1935) – McFarland was born in County Tyrone, Ireland. In 1871 he gained a first class BA in Mathematical Sciences and in 1872 an MA. He won a scholarship to St. John's College Cambridge where he graduated BA in 1876 and MA in 1879. He then taught at Repton School, Derbyshire, where he started a physics laboratory, made a special study of the telephone and formed a Natural History Society. His background impressed members of the provisional Council of Ormond College, Melbourne, who selected him as Foundation Principal. The College opened in 1881. He taught mathematics and physics himself and appointed five outstanding tutors, including Daniel McAlpine. In 1886 he was elected to the University of Melbourne Council, and in 1910 was appointed University Vice-Chancellor. In 1914 he relinquished his position at Ormond College, and in 1918 was appointed Chancellor of the University, which position he retained until 1933. He was widely respected. [*ADB* **10**:266–267].

McLennon, Ethel Irene (1891–1983) – McLennan was a staff member of the Botany School at Melbourne University from 1915, teaching mycology and plant pathology from the mid 1920 until her retirement in 1956. As Professor Ewart never invited McAlpine into the Botany School, or encouraged his staff to do so, she never met him [D. G. Parbery, Australasian *Plant Pathology* **18**:47–56, 1989; S.C. Ducker, in *200 Australian Women: A Redress Anthology*, Women's Redress Press Inc, Sydney, 1988].

Madden, Hon. Walter (1848–1925) – Madden was born in Ireland, educated in London, Marist College Beauchamp, France, and St Patrick's College, Melbourne, where the family settled in 1857. He was a jackeroo at Skipton Victoria for a while before signing on as a midshipman in Victoria's navy. There his study of marine survey led to his joining the survey department of the Victorian Lands Office in 1866, and 5 years later was the surveyor for the Horsham district. Returning to Melbourne in 1879 he established his own business as surveyor, land valuer and financial agent. The following year he was elected to parliament where he served until 1894. He strongly supported the interests of farmers, but his extremely conservative views on many issue prevented his being offered a ministry. [*ADB* **10**: 370–371].

Maiden, Joseph Henry FRS (1859–1925) – Born in London and educated in sciences at the Universities of Cambridge and London, he arrived on a visit to Sydney intending to stay 12 months. In 1881 he was appointed curator of the new Technological Museum in Sydney which, after its relocation to Ultimo in 1893, he turned into a centre of applied scientific research. Maiden was a botanist. He visited England in 1900, returning with 600 of Sir Joseph Banks' Australian plant collections. He became interested in the Australian flora and economic botany and in 1896 was appointed Director of the Sydney Botanic Gardens and NSW Government Botanist. He was a member, and served terms as president, of several learned societies, including the *NSW Linnaean Society* [*ADB* **10**:381–383].

Massee, George Edward (1850–1917) – Massee was a Yorkshireman from a farming background. He studied at Cambridge University. He was somewhat of an adventurer in his youth before taking up natural history and focusing on mycology. Between 1892 and 1895 he completed four volumes of *The British Fungous Flora.* Cooke's *Handbook of British Fungi* was more comprehensive. Massee, however, had wide interests in coprophilous fungi and in taxonomic organization. In 1895 he was elected the first President of the British Mycological Society. [G.C. Ainsworth, *Introduction to the History of Mycology*, 1976].

Masson, Professor Sir David Orme FRS (1858–1937) – Masson matriculated at the University of Edinburgh in 1877, studied chemistry and graduated B.Sc in 1880, then D.Sc. in 1884. He was appointed to the foundation chair in chemistry at

Melbourne University in 1887. In 1916, he was instrumental in the formation and establishment of CSIR, the forerunner of CSIRO. Masson and McAlpine's long friendship appears to have begun in Edinburgh. Masson retired in 1923. [*ADB* **10**: 432–434].

Meeking, E. – Meeking was Senior Fruit Inspector for the Victorian Department of Agriculture, who in 1916 replaced Mr Turner on the Advisory Committee.

Millardet, Pierre-Marie-Alexis (1838–1902) – Millardet studied in Heidelberg then in Freiberg, Germany, where he studied under de Bary. On his return to France he took out doctorates in medicine and science. In 1869 he accepted the post of Assistant Professor of Botany at the University of Strasbourg where he and de Bary were colleagues for a year or more. In 1872 he moved to Nancy where he investigated phylloxera of the vines. Then in 1876 he was appointed professor of botany at the University of Bordeaux and retired in 1899. While in Nancy he sought means of saving French vineyards from the ravages of the phylloxera epizootic. In about 1870 he imported native American grape vines resistant to the pest. This saved the French wine industry but introduced a highly destructive fungal parasite, downy mildew. Through serendipity, keen observation and his knowledge of science he overcame the new threat to the wine industry through the invention of Bordeaux mixture, the first generally useful fungicide. [G.C. Ainsworth, *Introduction to the History of Plant Pathology*, 1981; *Encycopaedia Britannica*]

Moore, Sir Newton James (1870–1936) – Moore was born in Fremantle, Western Australia. On leaving school he worked as a sales assistant for 2 years before being articled to Alexander Forrest, a surveyor, and later to Sir Andrew Forrest, Premier of Western Australia from 1890 to 1901. From 1886 to 1890 Moore travelled extensively throughout the state, mapping its forests and noting the potential of other resources. In 1903 he was a member of a Royal Commission on Forests. He became active in local politics and by 1904 was elected to the Western Australian Legislative Assembly. He was elected Premier of Western Australia in 1906, retiring as Premier and from politics in 1910, after which he served the state in other capacities. He represented the Western Australian Government in London as its Agent-General from 1911 until 1916, so was responsible for managing McAlpine's consignments of cool-store apples sent London on the *SS Somerset*. In May 1915 he took command, as lieutenant-colonel, of the depot for Australian and New Zealand troops at Weymouth in Dorset. By 1917 he held the temporary rank of brigadier-general in charge of Australian and New Zealand troops in Britain. [*ADB* **10**: 567–569].

Müller, Baron Sir Ferdinand Jakob Heinrich von (1825–1896) – von Müller was born in Schleswig-Holstein and graduated in pharmacy in 1846. His extensive collection of plants from that state brought him to the attention of Professor Nolte at the University of Kiel, where he gained a PhD. Because of poor health he was advised to seek a warmer climate, and when Dr J. Preiss, who had recently returned from Western Australia, told him about the wonderful flora there, he lost little time in migrating, arriving in Adelaide in December 1847. There he worked as an assistant chemist and during his spare time made excursions far into the countryside collecting plants. He wrote papers on his botanical discoveries, which he sent to learned societies in England and Germany and to newspapers in Adelaide. In 1853 he moved to Melbourne, where on the recommendation of Sir Joseph Hooker he was appointed Victorian Government Botanist, which position he held from 1853 until 1896. He drew all the available information on Australian botany together in a six-volume compilation. He became a great asset to his adopted land, to which he contributed for over 50 years. [*ADB* **5**:306–308].

Nitobe, Inazo (1862–1933) Nitobe was born in Japan. After a disappointing start to his tertiary education at Sapporo Agricultural College in Japan, he studied economics and political science at John Hopkins University in the USA for 3 years, In America he became involved with the Quakers, where he met his future wife, Mary Elkinton. Then in Germany he obtained a doctorate in agricultural economics, the first of five doctorates. After a stellar career in the Japanese public service, and following the establishment of the League of Nations in 1920, Nitobe was appointed Under-Secretary-General of the League and Director of the International Committee on Intellectual Cooperation which later became UNESCO under the UN's mandate. In 1921 he participated in the 13th World Congress of Esperanto, which was later proposed as the working language of the League of Nations. The French, however, vetoed the proposal, French being the diplomatic and international language of the time [S.M. Snipes, The Life of a Japanese Quaker Inazo Nitobe, *Friends Journal* 1st August, 2011].

Nobelius, Carl A. (1851–1921) – Nobelius was born in Finland to Swedish parents who were relatives of Alfred Nobel of Peace Prize fame. He migrated to Australia in 1872 and in 1886 purchased an initial 200 acres at Emerald where he established a highly successful, internationally respected shrub and tree nursery. By the turn of the century he employed 80 hands year-round and sold in the order of three million plants annually, worldwide. Most of the orchard site is now contained within Emerald Lake Park and his erstwhile packing shed is now a venue for Puffing Billy Railway functions [*ADB* **5**:343–344].

Norris, Dr W. Perrin – Norris was appointed the first Commonwealth Director of Quarantine within the Department of Trade and Customs. As such, he was Chairman of the Commonwealth Advisory Committee on Bitter Pit in 1911 and 1912. In September 1911, Norris travelled overseas, leaving Cumpston in charge of the federal service until May 1912. Then at the end of that year Norris returned to London, and Cumpston was appointed Director in May 1913. He was responsible for recommending the site for the Royal Canberra Hospital in 1912. [*Wikipedia*].

Parkes, Sir Henry (1815–1896) – Henry Parkes was born to tenant farmers on *Stoneleigh Abbey* Estate, Warwickshire, England. In 1823, debt forced the family off the estate. Over the following two decades, life was hard and Henry's education limited and imperfect. In 1838, he and his wife went to London to try their luck but survived only a few weeks by his pawning his tools. This decided them to leave for New South Wales as bounty migrants. They boarded the *Strathfieldsaye* on 27th March 1839, and reached Sydney 4 months later on 25th July. After early business failures he became a successful writer in spite of his patchy education. By 1850 he had joined the *Australian League* to work for universal suffrage and to transform the Australian Colonies into a Great Federal Republic. He won a seat in the NSW Legislative Council in 1854 and over the next 40 years worked hard to realise his dream of a federal system of government for Australia. It was a stormy time in the NSW colonial political era and over the next two decades Parkes was voted in and out of Parliament, and when in Parliament was in and out of office. He became Premier for some months in 1877 and again in 1878 but in 1881 took leave and a sea voyage to England to recover from poor health. On his return he continued his work toward federation. He was a politician of great stature, probity and skill as an administrator, highly respected, now regarded as the *Father of Federation* [ADB **5**: 399–406].

Pascoe, Hon. Thomas (1859–c.1933) – Pascoe was born near Clare in South Australia. He was a member of the S.A. Legislative Council from 1900 to 1933, Honorary Minister of Agriculture for S.A from 22nd December 1909 until 3rd March 1910, and Minister for Agriculture and Irrigation from 17th February 1912 until 3rd April 1915. [*South Australian Register*, Tuesday 22nd May 1900].

Peacock, Sir Alexander KCMG (1861–1933) – Peacock was born at Creswick in Victoria. He served as Premier of the State during 1901 and 1902, again from 1914 to 1917 and for a short third period in 1924. He was speaker of the Victorian Parliament from 1928 until his death [*ADB* **11**:173–175].

Pearson, Alfred N. – Pearson was Victorian Government Chemist in 1890. He was one of the Steering Committee of three appointed to draw up resolutions for the Rust-in-Wheat meeting of delegates from Victoria, South Australia, New South Wales and Queensland held in Melbourne in March 1890. He prepared the Minutes of Proceedings of the meeting. He and W. Farmer wrote the appendices to the Minutes [*Brisbane Courier Mail* 17-3-'90].

Pickering, Percival Spencer Umfreville, FRS (1858–1920) – Pickering was a chemist and horticulturalist. From 1881 he lectured at Bedford College until 1887 when he lost an eye in a laboratory explosion. From 1894 he was director of Woburn Experimental Fruit Farm established as a private enterprise by Pickering and the Duke of Bedford. [Obituary, *Biochemical Journal* **15**: 1–3, 1921].

Quinn, George (1870–c1945) – He was the South Australian Government's Horticultural Inspector & Chief Inspector Vegetation, also the Officer-in-Charge of the South Australian Experimental Orchard at Unley and Horticultural Instructor. From 1914 until 1935 he was Deputy Director of the Department of Agriculture, South Australia [*Encyclopaedia of Australian Science*].

Ramsay, Sir Andrew Crombie LLD, FRS (1814–1891) – Ramsay was educated at Saltcoats, and Glasgow in Scotland. He became a highly respected geologist and in 1863, having prepared a course of six lectures on geology for working men at the Royal School of Mines, he expanded them into a book, *The Physical Geology and Geography of Great Britain*, which was so well received it was republished for the 5th time in 1878. [*Oxford Dictionary of National Biography*].

Richardson, Professor Arnold Edwin Victor (1883–1949) – Richardson was born in Adelaide. In 1902, he obtained a first-class diploma from Roseworthy Agricultural College, and from 1904 to 1910 attended the University of Adelaide, gaining a B.A. (1907), a B.Sc. (1908) and an M.A. (1910). In 1908 he was transferred to the State Department of Agriculture as a plant breeder. In 1911, while acting Director of Agriculture, he was appointed Superintendent of Agriculture to the Victorian Government, establishing research stations at Werribee and Rutherglen, and laying the foundations of scientific agriculture in Victoria. In 1917 he had become a part-time lecturer in the faculty of agriculture and a member of the committee inquiring into the teaching of agriculture. In 1918 he investigated agricultural education and research in North America. From 1919 to 1924 he was first Dean of the reconstructed faculty and first Director of the School of Agriculture at Melbourne University. By 1920, he had reorganised the cereal industry in Victoria, and started a pasture improvement program. In 1924 he inaugurated the Better Farming Train to implement agricultural extension. That year, the University of Melbourne awarded him the degree of D.Sc. He then accepted appointment to the foundation Chair of Agriculture in the University of Adelaide and Director of the Waite Institute. [*ADB* **11**: 377–379].

Rothera, Arthur Cecil Hamel (1880–1915) – Rothera was a graduate of Cambridge University. In 1906 he was invited to the first lectureship in biochemistry in Australia at the Medical Faculty of the University of Melbourne. There he taught medical, pharmacy and agricultural science students, his work being of special value to scientific agriculture. He wrote a number of significant papers, his ketone test being still used for the diagnoses of diabetes. He enlisted with the rank of Captain in the First Australian Infantry Forces in 1915 and, while working among patients suffering meningitis, died within 2 weeks of enlisting. [Obituary, *Biochemistry Journal* **10**(1): 11–13, 1916; *Historical Records of Australian Science* **23**: 120–131, 2012].

Scott, R. – Scott was the Government Chemist attached to the Victorian Department of Agriculture in 1911 and a strong supporter of McAlpine.

Shoobridge, L. M. – Shoobridge was a leading Tasmanian orchardist who for some years was President of the National Fruit Growers; Association of Australasia. In the 1914 Federal elections he won a Senate seat for Tasmania (*The Barrier Miner* p. 1, 10-8-'14).

Sutherland, Alexander (1852–1902) – Sutherland was born in Glasgow, Scotland. In 1862 the family migrated to Sydney for reason of his father's health, and there Alexander trained as a teacher with the Council of Education. In 1870 the family moved to Melbourne where Alexander attended Melbourne University and between 1871 and 1875 gained B.A. and M.A. degrees. He was appointed mathematics master at Scotch College, then in 1877 headmaster of Carlton College, making it an immediate success. In 1892 he retired briefly to Dromana but because of the depression of 1892-'93 he returned to Melbourne and took up journalism. He was a member of the University of Melbourne Senate, and in 1901 became Registrar of the University of Melbourne (Blainey, 1957). From 1902 he lectured in English language and literature.

He was a man of many talents and much involved in public affairs [*ADB* **6**: 222–223; G. Blainey, *A Centenary History of the University of Melbourne*, 1957].

Swinburne, George (1861–1928) – Swinburne was born in England, became a highly educated practical engineer who arrived in Australia in 1886 where he joined his uncle and partner as a consultant gas and hydraulic engineer. He was elected to the Victorian Parliament in 1902 and was Minister for Agriculture from 1905 to 1908 [*ADB* **12**: 150–152].

Thiselton-Dyer, William Turner FRS, KCMG (1843–1928) – After graduating in Natural Science from Oxford University in 1867, he was appointed Professor of Natural History at the Royal Agricultural College in Cirencester, then Professor of Botany at the Royal College of Science for Ireland in Dublin. In 1872, he was appointed Professor at the Royal Horticultural Society in London on the recommendation of Joseph Dalton Hooker, whose daughter he later married. In 1875 he became Assistant-Director of the Royal Botanic Gardens at Kew and in 1905, following Hooker's retirement was appointed Director. He was a Fellow of the University of London and in 1880 Charles Darwin and George Bentham proposed him as a Fellow of the Royal Society. He held several exalted posts between 1887 and 1909 including botanical advisor to the Secretary of State for the Colonies from 1902 to 1906 [*Oxford Dictionary of National Biography*].

Thomson, Sir Charles Wyville FRSE, FRS, FLS, FGS, FZS (1830–1882) – A keen student of natural history, in 1851 he was lecturer in botany at the University of Aberdeen and between 1853 and 1870 he held Chairs of Natural History at Queen's College Cork in 1853, Geology at Queen's College Belfast in 1854, then the Chair of Botany and Zoology at Queen's Belfast in 1860. In 1868 he accepted the Chair of Botany at the Royal College of Science, Dublin, and in 1870 he was appointed Regius Professor of Natural History at the University of Edinburgh, where he died in March 1882. [*Encyclopaedia Britannica*].

Topp, Charles Alfred (1847–1932) – Topp was born in Huddersfield, Yorkshire, England. With his family he migrated to Melbourne in 1858, where he graduated M.A., LL.B from Melbourne University in 1869. His early career was in education. By 1884 he was senior inspector of schools for Victoria and later that year he was appointed Principal of the Teachers' Training College in Spring Street, which was relocated to Carlton as a residential college under his control in 1889. In 1890 he was Chairman of the Board of Public Health where he served with distinction over two decades. He was a keen naturalist, served a term as President of the Field Naturalists Club of Victoria and in 1887 was elected a Fellow of the Linnean Society of London, nominated by Sir Ferdinand von Müller [*ADB* **6**: 289–290].

Tryon, Henry (1856–1943) – Tryon was born in England. He was related to Lord Tryon and to Admiral Sir George Tryon and through the latter was connected to the family of author Anthony Trollope. He began his career as a medical student in London but was drawn into natural history. He then went plant collecting in Sweden and in New Zealand, where he managed a property for his father before settling in

Brisbane in 1882. There he established a reputation for developing an improved vaccine for pleuropneumonia, and cures for other livestock problems. He worked on rabbit control for the NSW Government during 1888-'89. From 1883 to 1893 he was appointed assistant curator of the Queensland Museum and in 1894 he was appointed Government Entomologist. In 1901 his designation was changed to Government Entomologist and Vegetable Pathologist. He was active in community affairs and widely involved in scientific societies. He was highly respected. Tryon was Queensland's representative on the Commonwealth Advisory Committee on Bitter Pit (*Queensland Naturalist* **31**: 4–12, 1991; *ADB* **6**: 305–306, *ADB* **12**: 272–273).

Tudor, The Hon. Francis Gwynne (1866–1922) – A hatter by trade, Tudor was elected to the Federal Parliament in 1901 and became an effective and highly respected Minister of Trade and Customs from 1908 to 1916. [*ADB* **12**: 281–182].

Turner, J.C. – Turner was the Chief Horticultural Officer of the Victorian Department of Agriculture, and Quarantine Officer for Plants under the Federal Quarantine Act. He was one of the Commonwealth Government's three representatives on the Commonwealth Advisory Committee on Bitter Pit until replaced by E. Meeking in 1916.

Urquart, Robert – Secretary of the Pharmacy Section of the School of Medicine, Edinburgh and lecturer in Materia Medica and Pharmacology in the School.

Ward, Harry Marshall (1854–1906) – During the 1890s until his death in 1906, Marshall Ward was considered the most influential plant pathologist in the United Kingdom. He had left school at 14, attended evening classes in South Kensington where his teachers included Thomas Huxley and William Thisleton-Dyer, then studied at Owens College Manchester and won a scholarship which took him to Christ's College, Cambridge. He then went to Würzburg to study with Anton de Bary and Julius Sachs, after which he returned to Kew as Thisleton-Dyer's assistant. He then went to Sri Lanka where his contribution to understanding the gravity of the coffee rust epiphytotic and helped drive the change from growing coffee to growing tea. It also initiated his interest in the physiology of host-parasite relationships. He finished his career as Professor of Botany at Cambridge University and is credited with modernising the teaching of plant pathology [*Annual Review of Phytopathology* **32**: 21–25, 1994; P.G. Ayres, *Harry Marshall Ward and the fungal wall of death*. Publisher, American Phytopathological Society, 2005].

Waterhouse, Professor Walter Lawry (1887–1969) – Waterhouse, who was born in Maitland NSW, graduated B.Sc.Agr. from Sydney University in 1914 with a first class degree and the University Medal. In 1915 he enlisted in the army, was commissioned, fought at Pozieres, was badly wounded and repatriated. He was awarded the Military Cross for conspicuous bravery. Supported by a Walter and Eliza Hall fellowship in 1918, he attended the Imperial College of Science and Technology, London, and obtained its Diploma in 1921. He returned to Australia via the USA where he met Professor E.C. Stakman, an authority on rusts. On his return in 1921, Waterhouse was appointed lecturer in agricultural botany at Sydney University, and in 1923 also lectured in plant pathology, genetics and plant breeding.

He produced important new wheat varieties resistant to rust and with improved baking qualities. He laid the foundation for breeding wheat at Sydney University, which is still strong today. He corresponded with McAlpine for advice and to obtain his results from tests of new barley genotypes for rust resistance. Waterhouse may have written an appreciation of both McAlpine and Cobb after their deaths in 1932, on behalf of the Sydney University Agriculture Graduates' Association in their journal *Suaga*. Waterhouse was also greatly admired by his students. [*ADB* **12**: 391–392].

Wedge, Erica [née McAlpine] (1895–1985) – Erica Wedge was the fourth of McAlpine's five daughters. After three of her sisters died in 1952 and the fourth in 1971, Wedge gathered the family papers and did what she could to preserve as much information as possible relating to her father's life and work. She did so in the hope that the difficulties he faced during the bitter pit investigation and the way he was treated at the end of it, would eventually be told. She was an honoured guest at the opening of the Plant Pathology Congress held in Melbourne in 1983.

Weste, Gretna Margaret AM (1917–2006) – Gretna Weste was born to Australian parents in Dumfries, Scotland not far from Gretna Green. The family returned to Australia in 1920 and on leaving school, Gretna studied botany at Melbourne University where Ethel McLennan was one of her lecturers. She was a brilliant student, graduated B.Sc. (Hons) in 1938, being awarded exhibitions and a Howitt Natural History Scholarship. On graduation she worked with the Victorian Forests Commission and CSIRO Forest Products. She stopped work in 1942 to raise a family. In 1955, because of her husband's ill health, she taught biology, proving to be a demanding but inspirational teacher. She joined the staff of the School of Botany at Melbourne University in 1960, where in addition to her teaching duties she studied the biology of the pathogen *Ophiobolus graminis*, the cause of *Take All* in wheat. She was proud of the fact that Professor S.D. Garrett, who had studied the same disease at the Waite Research Institute in South Australia some 30 years before, was one of her examiners. She graduated Ph.D. in 1968. She then turned her attention to the management of dieback in Australian forests caused by *Phytophthora cinnamomi* and made an enormous contribution to its control. During the early 1980s she chaired the organising committee for the Fourth International Congress of Plant Pathology held in Melbourne in 1983, and the same year was awarded a D.Sc. She received an Order of Australia in 1989. She was made an Honorary Member of the Australasian Plant Pathology Society in 1992, and elected President of the Society for 2005/2006. She was also Patron of the Australasian Mycological Society from 1999 until her death (Obituary: *Australian Mycologist* **25**: 4–12, 2007).

Whetzel, Professor Herbert Hice (c.1867–1944) – Whetzel was foundation Professor of Plant Pathology at the University of Cornell, Ithaca, NY, appointed in 1907. He was of German-Dutch heritage, born into a farming family in Indiana USA where he developed a deep interest in natural history. He was an early extension officer and a prolific writer of scientific articles. He was also a great admirer and friend of Daniel McAlpine, with whom he corresponded over many years [*Annual Review of Phytopathology* **18**: 27–36, 1980].

White-Haney, Rose Ethel Janet (Jean) (1877–1953) Jean White obtained a B.Sc. degree from Melbourne University in 1904, M.Sc. in 1906 and her D.Sc. in 1909. In 1910 she was appointed a research fellow working with Professor Ewart to investigate the enzymes and latent life of seeds. In mid-1910 the Victorian Government provided funds for research into the cause of bitter pit of apples. In 1912, the Queensland Board of Advice on Prickly Pear Destruction appointed her Officer-in-Charge of its experimental station at Dulacca 261 miles west of Brisbane, where she tested various insects and fungi as potential bio-control agents. By the time the Dulacca Station closed in 1916 she had also conducted over 10,000 tests on chemical poisons. She identified a moth, *Cactoblastis cactorum* as the most effective bio-agent, as well as the most practicable and economic means of control. It was introduced in 1926 with resounding success. She married V.W. Haney, an Agricultural Chemist in 1915. [*ADB* **16**: 616–617].

Willis, Dr James Hamlyn AM (1910–1995) – In 1930 Jim Willis gained a diploma from the Victorian School of Forestry, Creswick, and for the next 7 years worked as a forest officer with the Victorian Forests Commission. In October 1937 he joined the National Herbarium of Victoria as a taxonomic botanist, and commenced studies for a science degree at the University of Melbourne, graduating B.Sc. (Hons) and later D.Sc. Between 1958 and 1959 he was Australian Botanical Liaison Officer at the Royal Botanic Gardens, Kew, UK, and in 1961 was appointed Assistant Government Botanist for Victoria. Between 1970 and 1972 he was Acting Director of the Royal Botanic Gardens, Melbourne, retiring in January 1972. [Obituary: *Muelleria* **9**: 1–4]

Wilson, James Phillips – Wilson was Honorary Minister of Industry and Agriculture for South Australia from 3rd March 1910 until 17th December 1912.

Appendix 2

Text of the Draft Statement sent by Professor A. J. Ewart to Hon. George Graham regarding the Bitter Pit Investigation, which he wished to publish in daily newspapers.

Covering Letter

National Herbarium
South Yarra

4th August 1911

The Hon. G. Graham M.L.A., Minister for Agriculture

Dear Sir,

I should be glad to have your consent to publish the enclosed statement. You are probably aware that immediately Dr. White's paper on Bitter Pit was published, she wrote at my request to the Commonwealth authorities stating that if the continuance of her researches would interfere in any way with the arrangements to be made between the Commonwealth and States for the study of this problem, she would discontinue them. The fact that no attempt was made to defend the poisoning theory of Bitter Pit during the public discussion has naturally given rise to the idea that it is entirely abandoned. Hence a statement such as that enclosed will clear the air, and will make things easier for Mr. McAlpine when he is publishing the results of his first seasons work. I have recently been informed that the West Australian Health Department after treating apples with prussic acid found that they developed Bitter Pit and concluded that it was directly due to the poison used. Should you desire it I should be glad to shew you specimens of apples in which I have produced the artificial bitter pit by solutions containing one part poison to a million of water.

I am your obedient servant

Alfred J. Ewart
Government Botanist

Ewart's Draft Statement

The Problem of Bitter Pit

Now that the arrangements between the Commonwealth and the States for the appointment of a special investigator to study this disease are completed, it is possible to approach the problem from a purely scientific standpoint. The fact that Dr. White and myself have rigidly refrained from taking part in the public discussion of this problem or from making any defence of the poisoning theory of Bitter Pit, may possibly have led to the idea that the poisoning idea of Bitter Pit is abandoned. This is not the case. It is still the only theory which has any facts to support it, and these facts are as follows. Symptoms closely resembling the disease if not identical with it can be produced by the external application of various metallic and organic poisons. It appeared in one half of one orchard (46 trees) sprayed with arsenate of lead but not in the other unsprayed half, (34 trees). Dr. White was able to detect lead in the diseased tissues of Bitter Pit apples but not in the healthy tissue of the same apples. It has been objected that the arsenate of lead might have entered the Bitter Pits after they were formed. This objection is impossible to answer, but there is no reason why, if the poison can enter at one time it should not also enter at another. The disease was recognised by Dr. Cobb who was at the time our leading Plant pathologist, as a new and obscure disease, first noticed just prior to 1890, that is about the time when the use of poisonous sprays for various pests began to come into general use[a].

It was mainly on the basis of these facts[b] that what was expressly stated to be a tentative theory, was put forward by Dr White and as the previously accepted 'burst cell' theory of Pols-Evans had shown by my own experiments to be incorrect, there is no other theory supported by proved facts[b] in the field.

On the other hand, more than one orchardist whose opinions demand respect, considers that the disease was known at a much earlier date, long prior to the general adoption of spraying. If so however it must have been very much less prevalent that at present, since other-wise the attention of the specialists would have been drawn to it. In addition as the result of an energetic research carried out by the Victorian Department of Agriculture, six cases were discovered in which Bitter Pit appeared on apples, guaranteed not to have been sprayed during one or two seasons previously, and judging from the law of averages, probably still more cases would have been discovered if the search had been carried out at an earlier period of the season. Of these cases, four sets were collected and examined. Three were undoubtedly Bitter Pit, one was a little doubtful and in none of them could the State Analyst detect any traces of poisons. Other instances of apples affected by true Bitter Pit were received from trees stated the have been unsprayed during the preceding one or more seasons in Tasmania, South Australia and West Australia.

A conflict of evidence of this kind is at first sight bewildering. It occurred to me however that a partial explanation might possibly be, that apples are extremely sensitive to poisons and that the traces required to produce Bitter Pit might be so minute as to be incapable of detection by chemical analysis in a mass of organic matter such as that of an apple. The fact that some plants are peculiarly sensitive to poisons has long been known in Plant Physiology. For instance, a certain small water plant is able to detect the presence of 1 part of copper in 100,000,000 parts of water, or in an even greater dilution, and other similar though not so pronounced cases are known among plants.

Accordingly experiments were tried by testing apples with more and more dilute solutions of various poisons, chiefly of soluble salts of lead, copper and mercury. It was found that if care was taken to give the poison a possibility of entering the apple, the artificial Bitter Pit symptoms, could be produced including in some cases a slight but distinct recondensation of starch, by dilute poisons down to strengths of 1 in 200,000 in the case of lead nitrate and of 1 in 1,000,000 in the case of copper sulphate and mercury chloride (one gram of poison in one million cubic centimetres of water). In addition it was found that peculiar marks resembling a kind of superficial Bitter Pit can be produced by floating sound apples in a fairly strong solution (1 %) of formaldehyde. Yet 3 weeks after, these apples contained no trace of the poison possibly because in this case we were dealing with a volatile substance.

This line of work seems likely to afford strong support to the poison theory of Bitter Pit In fact the marvel appears to be not that apples sometimes develop Bitter Pit but that they are ever free of it. The reasons why it does not always appear in orchards sprayed with poisonous compounds are that these compounds are mostly very sparingly soluble, that the unbroken cuticle of sound apples is quite impermeable to poisons whether soluble or not and that they can only enter with difficulty through the minute breathing pores in the outer surface[c].

It must not however be imagined that even the complete establishment of the poisoning theory will mean that there is nothing more to be done upon the problem of Bitter Pit. Poisons might enter the apple directly from outside or from the soil through the root or organic poisons might even be manufactured by the plant itself. In the last case an empirical cure might be discovered before the chain of causation was fully established. At present the immediate problem is to discover whether for instance, 5 % of the cases of Bitter Pit in the field are due to spray poisons and 95 % to other causes, or whether the reverse is the case. The fact[d] that an artificial Bitter Pit resembling the natural one in some cases as far as the reappearance of or traces of starch can be produced by the external application of poisons is a very weighty one, and it is easy to show that pores are present in the surface of a sound ripe apple through which poisons can enter. Any one who doubts this need only float an apple on a ½% solution of sulphate of copper or of nitrate of lead for 2 or 3 days and then keep them for a week or two. This simple experiment is an admirable one to use in Agricultural High Schools to demonstrate the action of poisons on plants. The more dilute the poison or the more insoluble it is, the longer it takes to produce the artificial Bitter Pit. Possibly this is the reason for the late development of the natural Bitter Pit in apparently healthy apples in storage.

In conclusion I may be permitted to remark that a scientific theory is not disproved until the facts upon which it is based are also disproved or shown to be incorrectly applied, and that an apparent conflict between two sets of facts is often capable of a simple explanation, such as the recognition of some previously unknown relating factor. In any case, Dr. White's researches, together with the investigations made by the Victorian Department of Agriculture and the enquiries set on foot by Dr. Norris, the Federal Director of Quarantine, have placed the problem at least 1 and possibly 2 years in advance of where it would otherwise have been. More progress has in fact been made towards a complete knowledge of this disease during the past season, than during the previous 20 years that have elapsed since the disease first came to prominence[e].

Author's notes:

This statement has been transcribed with the spelling, grammar and punctuation used by Ewart.

There are a number of pencilled comments on the margin of the statement, made by either Dr Cameron or an advisor, which indicate that there was not a great deal of confidence placed in the scientific reliability of much of the statement. Sadly some are difficult to read. The sections in question are underlined and as far as it is possible to decipher the notations, the marginal notes are as follow: -

(a) questioned [?] permeability of diseased tissue.
(b) above are not proved facts.
(c) single word like 'crepi'.
(d) questioned [?] 'fact'.
(e) This closing sentence appears to be another attempt to discredit McAlpine!

Appendix 3

Daniel McAlpine Memorial Lecturers

Speakers and Topics

Lectures are delivered at each Biennial Conference of the APPS. Invitations to deliver the lectures are extended to eminent scientists in recognition of their significant contributions to Australasian Plant Pathology.

1976 – Dr. Lilian Fraser*, *Diseases of Citrus Trees in Australia – the First 100 Years.*
1978 – Dr. David Griffin, *Looking Ahead.*
1980 – Mr. John Walker, *Taxonomy, Specimens and Plant Diseases.*
1983+ – Prof. Richard Matthews, *Relationships between Plant Pathology and Molecular Biology.*
1985 – Prof. Bob McIntosh* & Dr Colin Wellings – *Wheat Rust Resistance – The Continuing Challenge.*
1987 – Prof. Allen Kerr*, *Agrobacterium: Pathogen, Genetic Engineer and Biological Control Agent.*
1989 – Dr. Albert Rovira, *Ecology, Epidemiology and Control of Take-All, Rhizoctonia and Cereal Cyst Nematode in Wheat.*
1991 – Mr. John Walker – *Plant, Diseases and Plant Pathologists in Australasia – a personal view.*
1993 – Dr. John Randles*, *Viruses, Viroids, and Virologists of Australasia.*
1995 – Dr. Ron Close*, *The Ever-changing Challenges of Plant Pathology.*
1997 – Prof. John Irwin, *Biology and Management of Phytophthora spp. Attacking Field-Crops in Australia.*
1999 – Dr. Dorothy Shaw*, *Bees and Fungi, with Special Reference to Certain Plant Pathogens.*
2001 – Dr. Alan Dube, *Long-Term Careers in Plant Pathology.*
2003 – Dr. Mike Wingfield, *Increasing Threat of Diseases to Exotic Plantation Forests in the Southern Hemisphere.*
2005 – Dr. Gretna Weste*, *A Long and Varied Fungal Foray.*
2007 – Dr. Graham Stirling*, *The impact of Farming Systems on Soilborne Diseases.*
2009 – Dr. Phillip Keane, *'Lessons from the Tropics', An Ongoing Journey of Discovery.*
2011 – Prof Lester Burgess*, *A Love Affair with Fusarium.*
2013 – Dr. Shaun Pennycook, *Fungal Names in Flux.*

* Those who held office as Presidents of the APPS.
+ No Conference was held in 1982, because of the planning and organisation required for the Fourth International Plant Pathology Congress, held in Melbourne in October 1983.

The manufacturer's authorised representative in the EU is Springer Nature Customer Service Centre GmbH, Europaplatz 3, 69115 Heidelberg, Germany. If you have any concerns regarding our products, please contact ProductSafety@springernature.com

Printed and bound by CPI Group (UK) Ltd, Croydon, CR0 4YY
25/09/2026
02222841-0005